SpringerBriefs in Physics

For further volumes:
http://www.springer.com/series/8902

Valeriy Astapenko

Interaction of Ultrashort Electromagnetic Pulses with Matter

Springer

Valeriy Astapenko
Moscow Institute of Physics
and Technology
Moscow
Russia

ISSN 2191-5423 ISSN 2191-5431 (eBook)
ISBN 978-3-642-35968-2 ISBN 978-3-642-35969-9 (eBook)
DOI 10.1007/978-3-642-35969-9
Springer Heidelberg New York Dordrecht London

Library of Congress Control Number: 2013931509

Printed on acid-free paper

Springer is part of Springer Science+Business Media (www.springer.com)

Preface

This book is devoted to the theory of interactions between ultrashort electromagnetic pulses (USPs) and matter, including both the classical and quantum cases. This is a hot topic in modern physics thanks to significant progress in generating and shaping USPs over a wide range of carrier frequencies.

Special attention is given to the peculiarities of the USP–matter interaction, namely, the phase dependence of photoexcitation and the nonlinear dependence of the total process probability upon the USP duration for one- and sub-cycle USPs.

One of the important items in the book is the derivation and example applications of a simple new formula which describes the total photoprocess probability under the action of USPs in the framework of perturbation theory. This formula expresses the total probability in terms of the cross-section of the process in a monochromatic field and the Fourier transform of the electric field strength. It describes the phenomenon in the situation when the standard approach based on the concept of radiation intensity and process rate becomes inadequate. The resulting expression can be considered as the analog of the Fermi golden rule in the physics of ultrafast electromagnetic phenomena. The formula is used to describe the photoexcitation by USPs of atoms and optical centers in solids and also the scattering of USPs on atoms and in plasmas.

A significant place is devoted to thc formalism of the optical Bloch vector, which gives a visual geometrical interpretation of radiative processes, making it possible to represent the time evolution of the quantum system under radiation through the rotation of a three-dimensional vector representing the state of the system. Strong field–matter interactions are treated using the Bloch formalism in a two-level approximation for USPs with variable characteristics, including chirped USPs.

The book is intended for a wide circle of readers, including students of the corresponding specialities, graduate students and scientists, university lecturers, and in fact anyone taking an interest in the physics of ultrafast electromagnetic interactions. It is hoped that representatives from each of the above groups will find useful material for their specific needs, whatever their scientific or pedagogic interests and qualifications.

Contents

Chapter 1
Oscillator in an Electromagnetic Field

The model of a harmonic oscillator is widely used in the most diverse fields of physics. This is connected first of all with the fact that, for small deviations from the equilibrium position x_0, the potential energy of the system $U(x)$ is approximately described by a quadratic x-coordinate dependence, and a quadratic dependence of the potential energy is a characteristic feature of a harmonic oscillator. In actual fact, in the equilibrium position the force f acting on a particle is equal to zero. Since the force is defined by the first derivative of the potential energy with respect to the coordinate ($f = -U'_x$), the linear term in the potential energy expansion in terms of the deviation from the equilibrium position $(x - x_0)$ is also equal to zero, and the quadratic dependence remains:

$$U(x \approx x_0) \cong U(x_0) + \frac{1}{2}\frac{d^2U}{dx^2}(x - x_0)^2. \tag{1.1}$$

Given that the choice of potential energy origin is arbitrary, the potential energy of the system at the equilibrium point may be assumed to be zero, so that $U(x_0) = 0$. Now, the total energy of a harmonic oscillator is equal to the sum of the kinetic and potential energies (for simplicity we consider a one-dimensional case):

$$\mathrm{E} = \frac{m\dot{x}^2}{2} + \frac{m\,\omega_0^2\,(x - x_0)^2}{2}, \tag{1.2}$$

where $\dot{x}$ is the velocity, m is the mass, and ω_0 is the eigenfrequency of the oscillator. Comparing the potential energy of the harmonic oscillator [the second summand on the right-hand side of (1.2)] with the right-hand side of the expansion (1.1), we find that

$$\omega_0 = \sqrt{U''_{x^2}(x_0)/m}, \tag{1.3}$$

where the double prime denotes the second derivative of the potential energy with respect to the coordinate. It is assumed that $U''_{x^2}(x_0) \neq 0$.

V. Astapenko, *Interaction of Ultrashort Electromagnetic Pulses with Matter*, SpringerBriefs in Physics, DOI: 10.1007/978-3-642-35969-9_1,

Thus for small deviations from the equilibrium position, any physical system can be treated approximately as a harmonic oscillator, the eigenfrequency of which is determined in the one-dimensional case by (1.3).

The harmonic oscillator model is widely used to study the physics of oscillation processes. It can also be applied to the description of quantum transitions of electrons between stationary states under the action of an electromagnetic field on various physical systems, using the spectroscopic conformity principle.

1.1 Harmonic Oscillator in a Monochromatic Field

We begin by considering an elementary case of interaction between a charged harmonic oscillator and the electric field of a monochromatic electromagnetic wave. Hereafter we will assume that the wavelength λ is much longer than the oscillator dimension a:

$$\lambda \gg a. \tag{1.4}$$

The inequality (1.4) is a precondition for applying the *dipole approximation*, which plays an important role in the physics of electromagnetic processes. It is easy to check that the relation (1.4) is true for a wide wavelength range if one considers the interaction of radiation with atomic particles, the characteristic size of which is defined by the Bohr radius: $a \approx a_B \cong 0.53$ Å (it will be recalled that 1 Å $= 10^{-8}$ cm). In the classical picture, the Bohr radius defines the size of the electron orbit nearest to the nucleus in a hydrogen atom.

When (1.4) is satisfied, the coordinate dependence of the electromagnetic field can be neglected, and for a monochromatic electric field strength, one can set

$$E(t) = E_0 \cos(\omega t + \varphi_0), \tag{1.5}$$

where E_0, ω, φ_0 are the amplitude, frequency, and initial phase of the electric field, respectively. A monochromatic field is characterized by the fact that its amplitude and initial phase are constants. The electric field amplitude determines the radiation intensity in vacuum averaged over the period $T = 2\pi/\omega$ according to the equation

$$I = c\,\frac{E_0^2}{8\pi}, \tag{1.6}$$

where c is the velocity of light in free space. It will be recalled that the intensity is the quantity of radiant energy passing through a unit area per unit time, usually measured in watts per square centimeter.

For the quantitative characterization of an electric field in atomic physics, the usual reference quantity is the atomic strength E_a, equal to the strength of the atomic field at the first Bohr orbit of a hydrogen atom:

$$E_a = \frac{m_e^2 e^5}{\hbar^4} \cong 5.14 \cdot 10^9 \text{ V/cm}, \tag{1.7}$$

where m_e is the electron mass. According to (1.6), the atomic radiation intensity corresponding to the atomic strength of the electric field (1.7) is

$$I_a = c \frac{E_a^2}{8\pi} = \frac{m_e^4 e^{10} c}{8\pi\hbar^8} \cong 3.52 \cdot 10^{16} \frac{\text{W}}{\text{cm}^2}. \tag{1.8}$$

The harmonic oscillator approximation can be used to describe the interaction between radiation and atomic particles for small enough electric field amplitudes and radiation intensities:

$$E \ll E_a, I \ll I_a. \tag{1.9}$$

When the inequalities (1.9) are satisfied, the influence of radiation on atomic electrons can be treated as a weak disturbance.

In the dipole approximation (1.4), the equation describing oscillations of the charged oscillator under the action of the electric field $E(t)$ has the form

$$\ddot{x} + 2\,\delta_0 \ddot{x} + \omega_0^2 x = \frac{e}{m} E(t), \tag{1.10}$$

where x is the deviation of the oscillator coordinate from the equilibrium position, while e, m, ω_0, δ_0 are the charge, mass, eigenfrequency, and damping constant of the oscillator, respectively, and dots denote time differentiation. The constant δ_0 determines the free oscillation decay time for the oscillator: $T_2 = 1/\delta_0$. The time T_2 is called the phase relaxation time or the transverse relaxation time. It plays an important role in the physics of electromagnetic interactions, determining in particular the width of the spectrum of radiation absorption by an ensemble of oscillators with the same eigenfrequencies (see Sect. 1.4.5).

It is not difficult to solve (1.10) by switching from time-dependent values to their Fourier transforms, that is, to frequency-dependent values. The Fourier transform of an arbitrary time function $f(t)$ is defined by

$$f(\omega) = \int_{-\infty}^{\infty} f(t) \exp(i\omega t)\, dt. \tag{1.11}$$

The benefit in using Fourier transforms comes from the fact that, after the transformation (1.11), time differentiation amounts to multiplication by the factor $-i\,\omega$, while integration is equivalent to division by this factor. As a result, the differential equation (1.10) becomes algebraic, and for the Fourier transform of the coordinate we obtain

$$x(\omega) = \frac{e}{m} \frac{E(\omega)}{\omega_0^2 - \omega^2 - 2i\omega\,\delta_0}. \tag{1.12}$$

The inverse Fourier transform of the right-hand side of (1.12), viz.,

$$f(t) = \int_{-\infty}^{\infty} f(\omega)\, \exp(-i\omega t)\, \frac{d\omega}{2\pi}, \tag{1.13}$$

gives the time dependence of the oscillator oscillations:

$$x(t) = \frac{e}{m} \int_{-\infty}^{\infty} \frac{E(\omega')\, \exp(-i\,\omega'\, t)}{\omega_0^2 - \omega'^2 - 2i\omega'\,\delta_0}\, \frac{d\omega'}{2\,\pi}. \tag{1.14}$$

Naturally, the integral in (1.14) depends on the explicit form of the function $E(\omega')$. For a monochromatic field the Fourier transform of the strength (1.5) is given by

$$E(\omega') = \pi\, E_0 \left[\exp(i\,\varphi_0)\, \delta(\omega' - \omega) + \exp(-i\,\varphi_0)\, \delta(\omega' + \omega)\right], \tag{1.15}$$

where $\delta(\omega)$ is the Dirac delta function. As can be seen from (1.15), in the case of a monochromatic field, the Fourier transform of the strength contains two frequencies $\omega' = \pm\omega$. It can be shown that the positive frequency corresponds to processes of radiation absorption, while the negative frequency is responsible for the radiation of electromagnetic waves.

Substituting (1.15) into (1.14) and carrying out elementary algebraic transformations results in the following expression for forced oscillations of the harmonic oscillator under the action of the monochromatic field:

$$x(t) = s(\omega)\, \cos(\omega\, t + \varphi_0) + q(\omega)\, \sin(\omega\, t + \varphi_0), \tag{1.16}$$

where the functions $s(\omega)$ and $q(\omega)$ are given by

$$s(\omega) = x_0 \frac{\omega_0^2\, (\omega_0^2 - \omega^2)}{\left(\omega_0^2 - \omega^2\right)^2 + 4\,(\omega\,\delta_0)^2} \approx \frac{x_0}{2}\, \frac{(\omega_0 - \omega)\, \omega_0}{(\omega_0 - \omega)^2 + \delta_0^2}, \tag{1.17}$$

$$q(\omega) = x_0\, \frac{2\,\omega\,\omega_0^2\,\delta_0}{\left(\omega_0^2 - \omega^2\right)^2 + 4\,(\omega\,\delta_0)^2} \approx \frac{x_0}{2}\, \frac{\delta_0\,\omega_0}{(\omega_0 - \omega)^2 + \delta_0^2}, \tag{1.18}$$

and $x_0 = e\,E_0 / m\,\omega_0^2$ is the amplitude of the oscillations of a free charge in a monochromatic field of amplitude E_0 and frequency ω_0. The approximate equations in (1.17)–(1.18) are valid for low radiation frequency detuning from the oscillator eigenfrequency, i.e., when

$$|\omega - \omega_0| \leq \delta_0 \ll \omega. \tag{1.19}$$

This a condition for resonance between the electric field and the harmonic oscillator, and the strong inequality here corresponds to an oscillator with low damping.

From the formula (1.16) it follows that the forced oscillations of the oscillator contain a component with the amplitude (1.17) that is *in phase* with respect to the electric field and a *quadrature* part with the amplitude (1.18) that is shifted by 90° with respect to the field phase (1.5). Furthermore, at the resonance frequency $\omega = \omega_0$, the in-phase amplitude is equal to zero and the quadrature amplitude takes its maximum value proportional to the ratio ω_0/δ_0.

Thus at the exact resonance and in the presence of damping, the forced oscillations of the harmonic oscillator are phase-shifted by 90° relative to the phase of the exciting monochromatic field. The oscillation amplitude in this case is inversely proportional to the damping constant of the oscillator and does not depend on the initial phase of the field φ_0.

In the case of forced oscillations of a *free* charge, when the eigenfrequency is equal to zero, (1.17)–(1.18) with neglected damping imply that $s = -e\,E_0/m\,\omega^2$, and $q = 0$, so the free charge (without damping) oscillates *in antiphase* with respect to the field causing the oscillations. Among other things, this explains reflection of electromagnetic waves from the surface of a conductor if their frequency is less than the plasma frequency characterizing collective properties of the electron density. Taking into account damping in the case of a free charge gives rise to a quadrature component proportional to the ratio δ_0/ω.

It should be emphasized that the quadrature amplitude of oscillations determines the average power $\langle P \rangle_T$ of energy exchange between the field and the oscillator over the electromagnetic wave period ($T = 2\pi/\omega$), defined as

$$\langle P \rangle_T = \frac{1}{T}\int_0^T e\,\ddot{x}(t)\,E(t)\,dt. \tag{1.20}$$

Substituting (1.16) into (1.20) and using $\left\langle \cos^2(\omega\,t + \varphi_0) \right\rangle_T = 1/2$, we find

$$\langle P \rangle_T = \frac{e\,E_0}{2}\,\omega\,q(\omega), \tag{1.21}$$

whence the power of energy exchange between the field and the oscillator is determined by the quadrature component of the forced oscillations (1.18). The contribution of the in-phase amplitude to the power disappears after averaging over the period of the electric field oscillation since $\langle \cos(\omega\,t + \varphi_0)$ $\sin(\omega\,t + \varphi_0) \rangle_T = 0$.

In the case under consideration, the power of electromagnetic interaction is positive ($\langle P \rangle_T > 0$), that is, radiant energy is *absorbed* by the oscillator and expended by losses due to damping of its oscillations.

The in-phase amplitude of the forced oscillations does not take part in energy exchange with radiation, but it nevertheless plays an important role since it defines the value of the refractive index of a substance. In the absence of damping ($\delta_0 = 0$), when the quadrature amplitude is equal to zero, the expression for the forced oscillations of the harmonic oscillator in the monochromatic field simplifies to

$$x(t, \delta_0 = 0) = \frac{e\,E_0}{m\left(\omega_0^2 - \omega^2\right)} \cos(\omega t + \varphi_0). \tag{1.22}$$

Hence it follows that, in the low-frequency range $\omega < \omega_0$, the oscillator oscillates in phase with respect to the electromagnetic wave field (1.5). As one goes through the resonance to frequencies $\omega > \omega_0$, the oscillator begins to oscillate in antiphase with respect to the external electric field. Connected with this is the appearance of negative values of the dielectric permittivity and magnetic permeability of a substance, a situation exploited to create meta materials with negative refraction.

Plots of the in-phase and quadrature amplitudes of oscillations of the harmonic oscillator in the monochromatic field are presented in Fig. 1.1.

Figure 1.1 shows that the quadrature amplitude is everywhere positive, while the in-phase amplitude changes sign as one passes through the resonance. With decreasing damping constant, the maximum values of the amplitudes grow and the widths of the maxima decrease. According to (1.18), the spectral width of the quadrature amplitude is δ_0. This means that the effective excitation of the harmonic oscillator by the monochromatic field occurs in the spectral range $(\omega_0 - \delta_0, \omega_0 + \delta_0)$.

The expression (1.14) can be transformed to a time integral if the order of the frequency and time integrations is swapped in the determination of the Fourier transform of the field. As a result, we obtain

$$x(t) = \frac{e}{m} \int\limits_{-\infty}^{\infty} dt'\; G(t - t')\, E(t'), \tag{1.23}$$

where

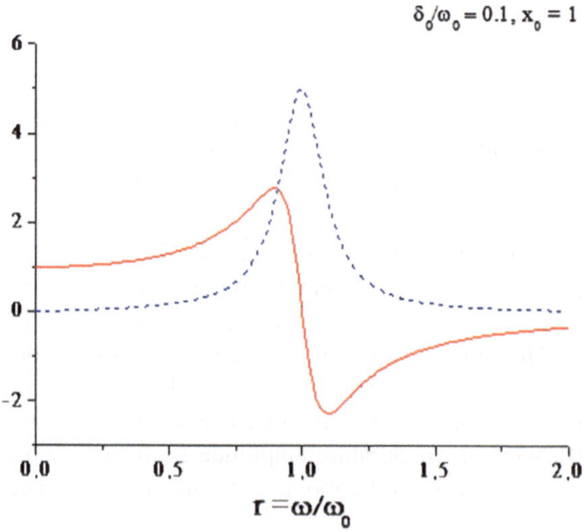

Fig. 1.1 The in-phase (*solid line*) and quadrature (*dotted line*) amplitudes of forced oscillations of the harmonic oscillator in a monochromatic field as functions of the frequency ratio $r = \omega/\omega_0$ for $\delta_0/\omega_0 = 0.1$ and $x_0 = 1$

$$G(\tau) = \int_{-\infty}^{\infty} \frac{\exp(-i\,\omega'\,\tau)}{\omega_0^2 - \omega'^2 - 2i\,\omega'\,\delta_0} \frac{d\omega'}{2\,\pi} \tag{1.24}$$

describes the harmonic oscillator response to the action of the electric field, also called the Green function. An explicit expression can be obtained for this function if the integral on the right-hand side of (1.24) is calculated by means of the residue theorem. The singularities of the integrand lie in the lower half-plane of the complex frequency ω'. Therefore the integral is equal to zero for $\tau < 0$, that is, for $t < t'$. This is a manifestation of the causality principle: the cause (electric field) cannot occur before the effect (oscillation of the oscillator coordinate). Calculation of the integral (1.24) gives

$$G(\tau) = \frac{\Theta(\tau)\,\exp(-\delta_0\,\tau)}{\sqrt{\omega_0^2 - \delta_0^2}} \sin\left[\sqrt{\omega_0^2 - \delta_0^2}\,\tau\right], \tag{1.25}$$

where $\Theta(\tau)$ is the Heaviside theta function (a unit "step"), equal to zero for $\tau < 0$ and unity for $\tau \geq 0$.

In the absence of damping the Green function (1.25) takes the rather simple form $G(\tau) = \Theta(\tau)\,\sin(\omega_0\,\tau)/\omega_0$.

Substituting the right-hand side of (1.25) into (1.23), we obtain

$$x(t) = \frac{e}{m\,\tilde{\omega}_0} \int_0^{\infty} \exp(-\delta_0\,\tau)\,\sin(\tilde{\omega}_0\,\tau)\,E(t-\tau)\,d\tau, \tag{1.26}$$

where $\tilde{\omega}_0 = \sqrt{\omega_0^2 - \delta_0^2}$ is the "renormalized" eigenfrequency that takes into account the oscillator damping. Physically, it is clear that the presence of damping should reduce the oscillator eigenfrequency. When writing down the right-hand side of (1.26), we went from integration with respect to the time t' to integration with respect to the time difference $\tau = t - t'$.

It should be emphasized that the explicit form of the electric field strength was not used to derive (1.26), so this formula is of a universal nature: it describes the response of the harmonic oscillator to arbitrary disturbances.

1.2 Harmonic Oscillator in a Pulsed Electromagnetic Field

1.2.1 Forced Oscillations

Let us consider the action of a Gaussian pulse on the oscillator, when the radiated electric field strength varies according to

$$E(t) = E_0 \exp\left(-t^2/\Delta t^2\right) \cos(\omega t + \Phi(t)), \tag{1.27}$$

where E_0 is the amplitude of the electric field strength, Δt is the pulse width, ω is the carrier frequency, and $\Phi(t)$ is the carrier phase with respect to the pulse envelope, which is in this case $E_{env}(t) = E_0 \exp\left(-t^2/\Delta t^2\right)$. It should be noted that the constant phase Φ is called the carrier envelope phase (CE phase). The case of a monochromatic field corresponds to an infinite pulse width $(\Delta t \to \infty)$ and $\Phi = \text{const}$.

To determine the coordinate position describing forced oscillations, (1.26) can be used. It is convenient to introduce a new variable $t' = t - \tau$. Then, applying the formula for the difference between two sine functions, we find

$$x(t) = \frac{e}{m\,\tilde{\omega}_0} \{\sin(\tilde{\omega}_0\, t)\, C(t) - \cos(\tilde{\omega}_0\, t)\, S(t)\}, \tag{1.28}$$

where

$$C(t) = e^{-\delta_0 t} \int_{-\infty}^{t} e^{\delta_0 t'} \cos(\tilde{\omega}_0\, t')\, E(t')\, dt', \tag{1.29}$$

$$S(t) = e^{-\delta_0 t} \int_{-\infty}^{t} e^{\delta_0 t'} \sin(\tilde{\omega}_0\, t')\, E(t')\, dt' \tag{1.30}$$

are the "cosine" and "sine"—the images of the electric field strength. Under the assumption $E(t \to -\infty) \to 0$, solution of (1.28)–(1.30) describes an oscillator that is not excited before the action of the field pulse: $x(t \to -\infty) \to 0$.

If the damping constant is greater than the reciprocal pulse width, then after cessation of the pulse action $t > \Delta t$, the amplitude of the oscillations of the oscillator goes to zero. Hereafter we will be interested in an opposite case, when damping is low $(\delta_0 \ll 1/\Delta t)$ and there is a time interval $\Delta t \ll t \ll 1/\delta_0$ during which the pulse has ceased, but oscillator damping has not yet been manifested. Considering such times in the formulas (1.28)–(1.30), the parameter δ_0 can be neglected and the integration can be extended to infinity. Then the values $C(\infty)$ and $S(\infty)$ are expressed in terms of the Fourier transform of electric field strength according to

$$C(\infty) = \text{Re}\{E(\omega' = \omega_0)\} \text{ and } S(\infty) = \text{Im}\{E(\omega' = \omega_0)\}. \tag{1.31}$$

In view of these relations, in the time interval $\Delta t \ll t \ll 1/\delta_0$, the formula (1.28) can be rewritten as

$$x(t) = \frac{e}{m\,\omega_0} |E(\omega_0)| \sin(\omega_0\, t - \arg[E(\omega_0)]), \tag{1.32}$$

where $|E(\omega_0)| = \sqrt{\{\mathrm{Re}(E(\omega_0))\}^2 + \{\mathrm{Im}(E(\omega_0))\}^2}$ is by definition the magnitude of the Fourier transform of the electric field strength at the eigenfrequency of the harmonic oscillator and $\arg[E(\omega_0)] = \mathrm{arctg}\{\mathrm{Im}(E(\omega_0))/\mathrm{Re}(E(\omega_0))\}$ is the argument of this Fourier transform.

Thus the formula (1.32) gives a compact representation of oscillations of the harmonic oscillator *after termination of an exciting pulse* of an electric field at times when oscillation damping can be neglected. For short, we will call such oscillations *asymptotic* oscillations. Since the field pulse has already ceased, it is natural that oscillations (1.32) occur at the eigenfrequency ω_0. From the expression (1.32) it follows in particular that an electric field pulse determines not only the amplitude, but also the phase of asymptotic oscillations. It will be shown in the following that under certain conditions this phase does not coincide with the initial phase of the electromagnetic pulse, but can differ noticeably from it.

1.2.2 Oscillator Excitation by Phase-Modulated Pulses

Let us consider two concrete examples of the Gaussian pulse (1.27) which are of interest from the point of view of applications to modern laser physics and chemistry (Fig. 1.2).

In the first case we assume that the carrier phase with respect to the envelope (the CE phase) is constant: $\Phi(t) = \varphi_0 = \mathrm{const}$, but that its value can be changed in a predetermined manner from one pulse to the next. The control of light-induced processes by changing the CE phase is called *phase control*. Phase control can now be achieved in practice. Indeed, fairly accurate methods of CE phase control have now been developed. In this case pulse widths can be very small, of the order of the period of oscillation at the carrier frequency:

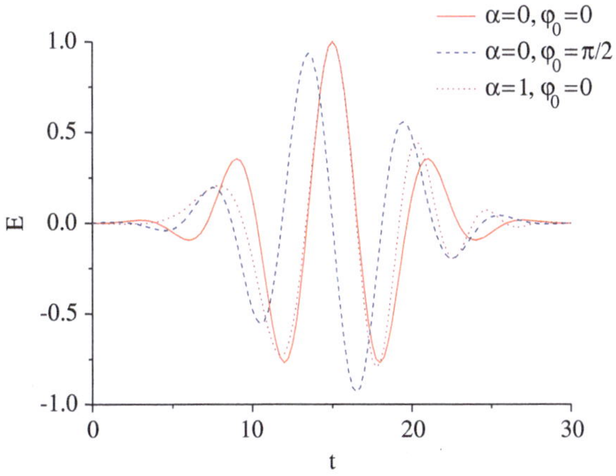

Fig. 1.2 Electric field pulses with different CE phases and chirps

$$\Delta t = n_p T = \frac{2\pi}{\omega} n_p, \tag{1.33}$$

where n_p is a number of order unity. Pulses satisfying this condition are called *ultrashort* pulses. If $n_p = 1$, a pulse is called a single-cycle pulse, and if $n_p < 1$, it is called a subcycle pulse.

It should be noted that, for subcycle pulses, (1.27) is replaced by other expressions for the electric field strength in the literature. We present here the expression from [1]:

$$E(t) = \mathrm{Re}\left\{ -i E_0 \left[\frac{\left(1 + i\frac{2t}{\omega \Delta t^2}\right)^2 + \frac{2}{\omega^2 \Delta t^2}}{1 + \frac{2}{\omega^2 \Delta t^2}} \right] \exp(-t^2/\Delta t^2) \exp(i\omega t + \varphi_0) \right\}. \tag{1.27a}$$

It should be noted that the conventional electric field pulse (1.27) and the modified pulse (1.27a) give the same results for $n_p \geq 2$ [1].

Single-cycle Gaussian pulses with different CE phases are presented in Fig. 1.2 for $\varphi_0 = 0$ (a cosine pulse) and $\varphi_0 = \pi/2$ (a sine pulse). It can be seen that the change in the CE phase influences the pulse shape.

In the second case of interest to us, $\Phi(t) = \kappa t^2$ holds. Such a pulse is called a *chirped* pulse, and the parameter κ is the so-called *time chirp*. In the chirped pulse under consideration the carrier frequency depends linearly on time: $\omega_c = \omega + \kappa t$. Chirped pulses find wide application in laser equipment operating at the femtosecond and attosecond time scales.

A single-cycle pulse with time chirp $\kappa = \Delta t^{-2}$ is shown in Fig. 1.2.

It follows from (1.32) that, in order to determine the asymptotic oscillations of the harmonic oscillator after the action of an exciting pulse, one must know the Fourier transform of the electric field strength in this pulse. In the first case under consideration ($\Phi(t) = \varphi_0 = \text{const}$), the Fourier transform of the field is

$$E(\omega') = E_0 \frac{\sqrt{\pi}}{2} \Delta t \left\{ \exp\left[-i\varphi_0 - \frac{(\omega - \omega')^2 \Delta t^2}{4} \right] + \exp\left[i\varphi_0 - \frac{(\omega + \omega')^2 \Delta t^2}{4} \right] \right\}. \tag{1.34}$$

For the pulse given by (1.27a),

$$\begin{aligned} E(\omega') = E_0 \frac{\sqrt{\pi}}{2} \Delta t \left[i \frac{\omega'^2 \Delta t^2}{2 + \omega^2 \Delta t^2} \right] \left\{ \exp\left[-i\varphi_0 - \frac{(\omega - \omega')^2 \Delta t^2}{4} \right] \right. \\ \left. - \exp\left[i\varphi_0 - \frac{(\omega + \omega')^2 \Delta t^2}{4} \right] \right\}. \end{aligned} \tag{1.34a}$$

One can see that the difference between (1.34) and (1.34a) consists in the first factor in square brackets and the sign of the last term in the brace.

The Fourier transform (1.34a) does not contain the constant component of the electric field strength ($E(\omega' = 0) = 0$) and satisfies some additional requirements for subcycle pulses (see [1]).

For chirped pulses, when $\Phi(t) = \kappa t^2$, one has

$$E(\omega') = \frac{\sqrt{\pi}\, E_0\, \Delta t}{\sqrt[4]{1+\alpha^2}} \exp\left\{ -\frac{\omega^2 + \omega'^2 + 2\, i\, \alpha\, \omega\, \omega'}{\Delta\omega^2} \right\} \cos\left\{ \frac{1}{2} \operatorname{arctg}(\alpha) - \frac{\alpha\,(\omega^2 + \omega'^2) - 2\, i\, \omega\, \omega'}{\Delta\omega^2} \right\}, \qquad (1.35)$$

where $\alpha = \kappa\, \Delta t^2$ is the dimensionless chirp and $\Delta\omega = 2\sqrt{1+\alpha^2}/\Delta t$ is the chirped pulse spectrum width.

Equations (1.34)–(1.35) are obtained by substituting (1.27) into the formula for the Fourier transform (1.11) and using tabulated values of the relevant integrals.

Substituting the Fourier transform of the electric field (1.34) into (1.32), we obtain the following expression for the coordinate of the harmonic oscillator after cessation of action of a short ($\Delta t \ll \delta_0^{-1}$) electromagnetic radiation pulse:

$$x(t) = \sqrt{\pi} \frac{e\, E_0\, \Delta t}{2\, m\, \omega_0} \sqrt{G(\omega_0,\, \omega,\, \Delta t)\, (1 + K(\omega_0,\, \omega,\, \Delta t)\, \cos(2\, \varphi_0))}\, \sin(\omega_0\, t + \psi), \qquad (1.36)$$

where

$$G(\omega_0, \omega, \Delta t) = \exp\left[-\frac{\Delta t^2\, (\omega_0 - \omega)^2}{2} \right] + \exp\left[-\frac{\Delta t^2\, (\omega_0 + \omega)^2}{2} \right] \qquad (1.37)$$

is the spectral line of excitation of the harmonic oscillator,

$$K(\omega_0,\, \omega,\, \Delta t) = \operatorname{sech}\left(\omega_0\, \omega\, \Delta t^2 \right) \qquad (1.38)$$

is the oscillation amplitude phase modulation factor [3], and

$$\psi(\omega_0,\, \omega,\, \Delta t) = \operatorname{arctg}\left\{ \operatorname{th}\left(\omega_0\, \omega\, \Delta t^2/2 \right) \operatorname{tg}(\varphi_0) \right\} \qquad (1.39)$$

is the initial phase of oscillations of the oscillator after the pulse ceases to act.

It is of interest that the oscillation amplitude phase modulation factor for the pulse (1.27a) is given *by the same expression* (1.38) obtained for the conventional pulse (1.27).

The spectral line of excitation of the harmonic oscillator by the pulse (1.27a) has the form

$$G_s(\omega_0,\ \omega,\ \Delta t) = \left[\frac{\omega_0^2 \Delta t^2}{2 + \omega^2 \Delta t^2}\right]^2 \exp\left[-\frac{\Delta t^2\,(\omega_0 - \omega)^2}{2}\right] + \exp\left[-\frac{\Delta t^2\,(\omega_0 + \omega)^2}{2}\right] \tag{1.37a}$$

rather than the analogous formula (1.37) obtained for the conventional electric pulse (1.27).

The phase modulation factor (1.38) describes the dependence of the oscillation amplitude of the oscillator on the CE phase (φ_0) of the electromagnetic pulse. For not too short pulses ($n_p \geq 1$), the condition $\omega_0\,\omega\,\Delta t^2 \gg 1$ is satisfied, when $K(\omega_0,\ \omega,\ \Delta t) \cong 0$, and phase modulation is absent. Figure 1.3 shows the dependence of the phase modulation factor on the dimensionless pulse width $\eta = \omega_0\,\Delta t$ for different ratios of the carrier frequency to the oscillator eigenfrequency $r = \omega/\omega_0$.

Figure 1.3 shows that the phase modulation factor has an appreciable value only for subcycle pulses, beginning with a half-cycle pulse, when $\eta < 3$, and also that the value $K(\omega_0,\ \omega,\ \Delta t)$ grows with decreasing carrier frequency.

Plots of the spectral line of excitation of the harmonic oscillator calculated using (1.37) and (1.37a) are presented in Fig. 1.4 for different electromagnetic pulse durations. From the given dependences it follows that the excitation line broadens with decreasing pulse width. For long pulses, the spectral line of excitation has the usual form of a Gaussian bell-shaped curve with a maximum at the oscillator eigenfrequency for both exciting pulses. There is also a blue shift of the spectral maximum due to excitation by the pulse (1.27a).

It should be noted that, for short multicycle pulses with controlled CE phase satisfying $\delta_0 \ll 1/\Delta t$, the width of the spectral line of excitation of the harmonic oscillator is determined by the pulse width and is equal to $\Delta\omega = 2/\Delta t$. It will be recalled that, in the monochromatic case, an analogous value is determined by the damping constant of the oscillator δ_0.

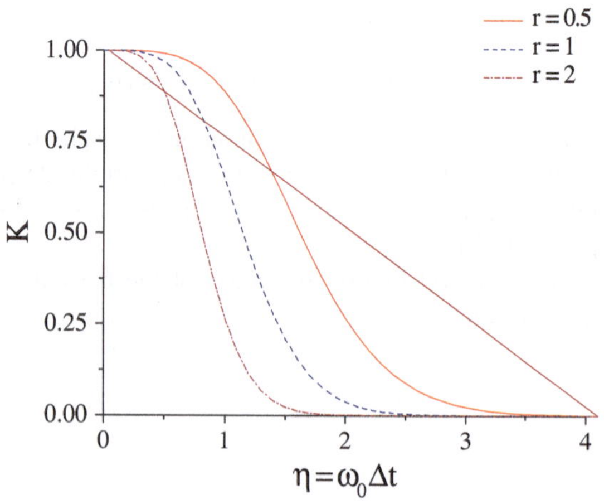

Fig. 1.3 Dependence of the modulation factor on the dimensionless parameter $\eta = \omega_0\,\Delta t$ for different values of the ratio $r = \omega/\omega_0$ under excitation by a pulse with the controlled CE phase

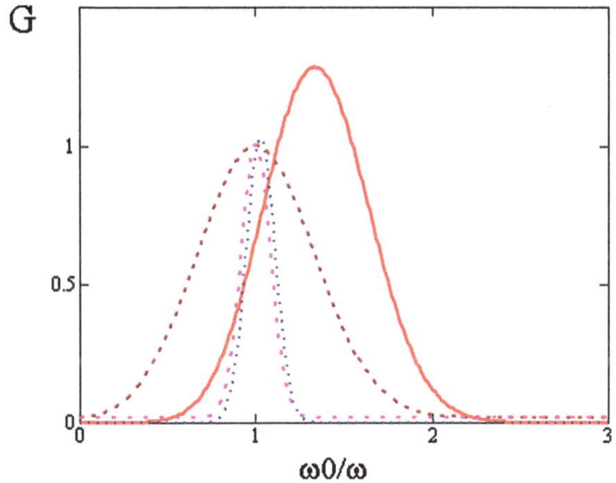

Fig. 1.4 The spectral function of excitation of the harmonic oscillator by pulses (1.27), (1.27a) for different pulse durations. *Solid line* half-cycle pulse (1.27a). *Dashed line* half-cycle pulse (1.27). *Dotted line* two-cycle pulse (1.27a). *Dash-dotted line* two-cycle pulse (1.27)

When the harmonic oscillator is excited by a chirped pulse, the formula for asymptotic oscillations has a structure similar to (1.36), except that, instead of the CE phase, the dimensionless chirp $\alpha = \kappa\,\Delta t^2$ will appear as control factor. The corresponding expressions are:

$$\tilde{x}(t) = \sqrt{\pi}\frac{e\,E_0}{2\,m\,\omega_0\,\sqrt[4]{1+\alpha^2}}\sqrt{\tilde{G}\left(1+\tilde{K}\,\cos f(\alpha)\right)}\,\sin\left(\omega_0\,t+\tilde{\psi}\right), \tag{1.40}$$

$$\tilde{G} = \exp\left\{-\frac{(\omega-\omega_0)^2}{\Delta\omega^2}\right\} + \exp\left\{-\frac{(\omega+\omega_0)^2}{\Delta\omega^2}\right\}, \tag{1.41}$$

$$\tilde{K} = \operatorname{sech}\left\{\frac{4\,\omega\,\omega_0}{\Delta\omega^2}\right\}, \tag{1.42}$$

$$f(\alpha) = \frac{2\alpha\,\left(\omega_0^2+\omega^2\right)}{\Delta\omega^2} - \operatorname{arctg}(\alpha), \tag{1.43}$$

$$\tilde{\psi} = \operatorname{arctg}\left\{\frac{\operatorname{tg}\left[\frac{\alpha\left(\omega_0^2+\omega^2\right)}{\Delta\omega^2} - \frac{1}{2}\operatorname{arctg}\alpha\right]\operatorname{th}\left[\frac{2\,\omega\,\omega_0}{\Delta\omega^2}\right] - \operatorname{tg}\left[\frac{2\,\omega\,\omega_0}{\Delta\omega^2}\right]}{1+\operatorname{tg}\left[\frac{\alpha\left(\omega_0^2+\omega^2\right)}{\Delta\omega^2} - \frac{1}{2}\operatorname{arctg}\alpha\right]\operatorname{th}\left[\frac{2\,\omega\,\omega_0}{\Delta\omega^2}\right]\operatorname{tg}\left[\frac{2\,\omega\,\omega_0}{\Delta\omega^2}\right]}\right\}, \tag{1.44}$$

where $\Delta\omega = 2\sqrt{1+\alpha^2}/\Delta t$ is the chirped pulse spectrum width.

Equation (1.41) implies that the pulse spectrum width will also determine the width of the spectral line of excitation of the harmonic oscillator. Since this width grows with the parameter $\alpha = \kappa\,\Delta t^2$, the coefficient $\tilde{K}$ will be somewhat greater than its CE analog (1.38) for the same pulse widths. Besides, the chirp dependence, though rather weak, is in the pre radical multiplier of the formula (1.40).

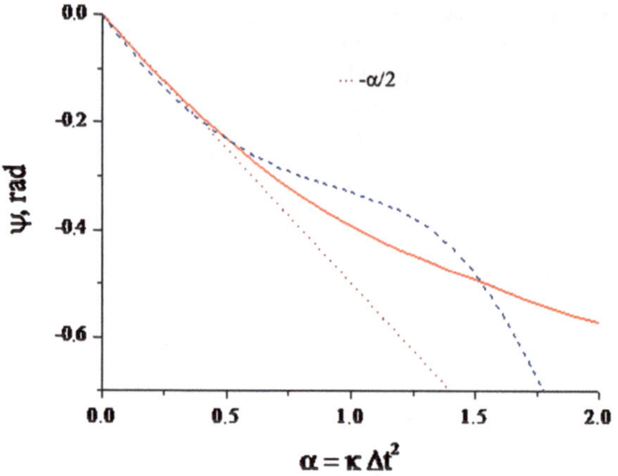

Fig. 1.5 Dependence of the phase shifts of asymptotic oscillations of the harmonic oscillator on the chirp value for different exciting pulse widths. *Solid line* one-cycle pulse. *Dashed line* half-cycle pulse

In the limit of a long pulse, when $\Delta\omega \ll \omega$ and hence $\mathrm{th}\left[2\,\omega\,\omega_0/\Delta\omega^2\right] \cong 1$, (1.44) simplifies to

$$\tilde{\psi} = \frac{\alpha\left(\omega_0^2 - \omega^2\right)}{\Delta\omega^2} - \frac{1}{2}\,\mathrm{arctg}\alpha, \tag{1.45}$$

so at the resonance we then have a simple relation for the shift in the harmonic oscillator phase $\tilde{\psi} = -(1/2)\,\mathrm{arctg}\alpha$. For low chirps $(|\alpha| < 1)$, it thus follows that the phase shift resulting from the action of a chirped pulse on the harmonic oscillator is proportional to the dimensionless chirp $\tilde{\psi} \approx -\alpha/2$. Figure 1.5 shows the shift in the phases of asymptotic oscillations as a function of the chirp value.

From Fig. 1.5 it follows that the distinction between the phase shift of (1.44) and the simple dependence $\tilde{\psi} \approx -\alpha/2$ is essential only for subcycle pulses and chirp values of greater magnitude.

1.2.3 Harmonic Oscillator Radiation

In the time interval $\Delta t \ll t \ll 1/\delta_0$ when the exciting pulse has already ceased and oscillation damping has not yet set in, the charged harmonic oscillator will emit electromagnetic waves according to the laws of electrodynamics.

The instantaneous power of dipole radiation is determined by the acceleration of the charged particle according to the well known formula

$$Q = \frac{2\,e^2\,\ddot{x}^2}{3\,c^3}, \tag{1.46}$$

where $\ddot{x}$ is the charge acceleration. This formula is true in the dipole approximation, when the inequation (1.4) holds, so the value (1.46) is called the dipole radiation power.

Of interest for practical purposes is the average radiation power over the oscillation period $\langle Q\rangle_T \equiv \langle Q\rangle$. In the case under consideration for the harmonic oscillator, this power can be expressed in terms of the total energy E of the oscillator by using the equation of motion, according to which $\ddot{x} = -\omega_0^2 x$. Hence we find $\ddot{x}^2 = \omega_0^4 x^2$. On the other hand, $\omega_0^2 x^2 = 2\,U/m$, where U is the potential energy of the harmonic oscillator, so $\ddot{x}^2 = 2\omega_0^2\,U/m$ and $Q = 4\,e^2\,\omega_0^2\,U/\,3\,m\,c^3$. Averaging the last equation over the period and taking into account the fact that $2\,\langle U\rangle_T = \mathrm{E}$, we obtain the required representation:

$$\langle Q\rangle = A_{sp}\,\mathrm{E}, \tag{1.47}$$

where

$$A_{sp} = \frac{2\,e^2\omega_0^2}{3\,m\,c^3} \tag{1.48}$$

is a coefficient with physical dimensions of reciprocal time that describes the probability of spontaneous emission, that is, emission arising in the process of free oscillations of a charged particle.

Emission of electromagnetic waves results in a loss of energy by the harmonic oscillator and in damping of its oscillations. Thus, if there are no other sources of losses, the coefficient (1.48) is equal to the damping constant δ_0 appearing in the equation of motion (1.10). It will be shown below that the parameter A_{sp} coincides with the Einstein coefficient for spontaneous radiation. In the optical range, when $\omega_0 \approx 10^{15}$ to $10^{16}\ \mathrm{s}^{-1}$, (1.48) implies a numerical value of $10^8\ \mathrm{s}^{-1}$ for the coefficient A_{sp}. This value represents a characteristic probability per unit time for spontaneous radiation in a transition of an atomic electron from one orbit to another.

The ratio of the spontaneous radiation coefficient (1.48) to the eigenfrequency can be written $A_{sp}/\omega_0 \approx r_e/\lambda_0$, where

$$r_e = \frac{e^2}{m_e\,c^2} \approx 2.8 \cdot 10^{-13}\ \mathrm{cm} \tag{1.49}$$

is the classical electron radius and $\lambda_0 = 2\,\pi\,c/\omega_0$ is the wavelength at the oscillator eigenfrequency. So when $A_{sp} = \delta_0$, the condition for weak damping of oscillations of the harmonic oscillator, viz., $\delta_0/\omega_0 \approx r_e/\lambda_0 \ll 1$, is true down to very short wavelengths, corresponding to the gamma range of light quantum energies.

The total energy of the oscillator is related to the amplitude of its free oscillations x_{amp} according to $\mathrm{E} = m\,\omega_0^2\,x_{amp}^2\Big/2$, obtained after averaging the squared coordinate over the period. In view of this relation, the average radiation power over the period is

$$\langle Q\rangle = \frac{2\,e^2\omega_0^4}{3\,c^3}\,x_{amp}^2. \tag{1.50}$$

Determining the amplitude of the asymptotic oscillations x_{amp} from (1.36) and using (1.50), we obtain the average radiation power for the harmonic oscillator after its excitation by a Gaussian pulse with controlled CE phase [2]:

$$\langle Q\rangle = \frac{\pi}{6}\frac{e^4}{m c^3}(\omega_0\,\Delta t)^2 E_0^2\, G\,(1+K\,\cos(2\,\varphi_0)), \tag{1.51}$$

where the spectral line of excitation of the oscillator G and the phase modulation factor K are determined by (1.37) and (1.38).

The expression (1.51) indicates that by changing the CE phase it is possible to control the charged oscillator radiation power after cessation of action of a pulse, provided the factor K has an appreciable value. In the case under consideration this is possible only for subcycle exciting pulses, according to the formula (1.38) and the plots of Fig. 1.3.

For excitation of the harmonic oscillator by a chirped pulse, instead of (1.51), a similar formula holds for its radiation power:

$$\langle\tilde{Q}\rangle = \frac{\pi}{6}\frac{e^4}{m c^3}\frac{(\omega_0\,\Delta t)^2}{\sqrt{1+\alpha^2}} E_0^2\,\tilde{G}(\alpha)\left(1+\tilde{K}(\alpha)\,\cos(f(\alpha))\right), \tag{1.52}$$

where $\alpha = \kappa\,\Delta t^2$ is the dimensionless frequency chirp and the functions $\tilde{G}$, $\tilde{K}$ and f are determined by (1.41)–(1.43). In this case the phase parameter (chirp) is included (through the pulse spectrum width $\Delta\omega = 2\sqrt{1+\alpha^2}/\Delta t$) in the determination of the spectral line of excitation of the oscillator (1.41), the factor $\tilde{K}(\alpha)$, and the common factor in the power expression (1.52). Therefore the radiation power of the oscillator excited by a chirped pulse can also be controlled for multicycle pulses, though with low efficiency.

1.3 Harmonic Oscillator in a Thermal Radiation Field

In thermal radiation, the amplitudes, phases, and polarizations of the electromagnetic field vary with time in a random manner. However, radiation averages, such as the spatial energy density u, are constant in time and space. It can be said that thermal radiation belongs to the class of random stationary and spatially uniform fields. The spectral density of stationary random field energy per unit volume is

$$\rho(\omega) = \frac{1}{(2\,\pi)^2}\int_{-\infty}^{\infty}\exp(i\,\omega\,\tau)\,\langle\mathbf{E}(t)\,\mathbf{E}(t+\tau)\rangle\,d\tau, \tag{1.53}$$

where angle brackets designate averaging over the field state and $\langle\mathbf{E}(t)\,\mathbf{E}(t+\tau)\rangle$ is the electric field strength autocorrelator which, under the assumption of stationarity, does not depend on the instant of time t. The formula (1.53) can be obtained from the expression for the spatial density of field energy $u = \langle\mathbf{E}^2\rangle/4\,\pi$

in an electromagnetic wave in vacuum by expanding into the Fourier integral and comparing with the equation $u = \int_0^\infty \rho(\omega)\, d\omega$ (see Appendix I).

Let us derive a useful relation between the average of the product of the Fourier transforms of electric field strengths and the spectral density of radiant energy. The relation has the form

$$\langle \mathbf{E}(\omega)\, \mathbf{E}(-\omega') \rangle = (2\pi)^3\, \rho(\omega)\, \delta(\omega - \omega'). \tag{1.54}$$

To prove this, we substitute the expressions for the Fourier transforms of strengths (1.11) into the left-hand side of this equation:

$$\left\langle \int_{-\infty}^{\infty} \exp(i\omega t)\, \mathbf{E}(t)\, dt \int_{-\infty}^{\infty} \exp(-i\omega' t')\, \mathbf{E}(t')\, dt' \right\rangle$$
$$= \int_{-\infty}^{\infty} \int_{-\infty}^{\infty} \exp(i\omega t - i\omega' t') \langle \mathbf{E}(t)\, \mathbf{E}(t') \rangle\, dt\, dt'.$$

We now introduce a new integration variable $\tau = t - t'$ and take into account the fact that the stationarity condition implies $\langle \mathbf{E}(t)\, \mathbf{E}(t') \rangle = \langle \mathbf{E}(t' + \tau)\, \mathbf{E}(t') \rangle = \langle \mathbf{E}(\tau)\, \mathbf{E}(0) \rangle$. Then we have

$$\int_{-\infty}^{\infty} \int_{-\infty}^{\infty} \exp(i\omega t - i\omega' t') \langle \mathbf{E}(t)\, \mathbf{E}(t') \rangle\, dt\, dt'$$
$$= \int_{-\infty}^{\infty} \int_{-\infty}^{\infty} \exp(i\omega (t' + \tau) - i\omega' t') \langle \mathbf{E}(\tau)\, \mathbf{E}(0) \rangle\, d\tau\, dt'$$
$$= 2\pi\, \delta(\omega - \omega') \int_{-\infty}^{\infty} \exp(i\omega\tau)\, \langle \mathbf{E}(\tau)\, \mathbf{E}(0) \rangle\, d\tau = (2\pi)^3\, \delta(\omega - \omega')\, \rho(\omega) \tag{1.55}$$

To obtain the last equation, we used the expression (1.53) for the spectral density of the electromagnetic field energy and the integral representation

$$\delta(\alpha) = \frac{1}{2\pi} \int_{-\infty}^{\infty} \exp(i\alpha t)\, dt \tag{1.56}$$

for the delta function. This proves (1.54).

Using (1.54), we can obtain the important relation between the power absorbed by the harmonic oscillator under the action of electromagnetic radiation and the spectral density of radiant energy (1.53). For this purpose we use the expression for the instantaneous power of interaction between the electric field $\mathbf{E}(t)$ and a charge e moving along the x axis, with the one-dimensional oscillator in mind:

$$P(t) = e\,\dot{x}(t)\,E_x(t). \tag{1.57}$$

For the harmonic oscillator, the coordinate position of forced oscillations $x(t)$ is given by (1.14), and it is easy to find the velocity $\dot{x}(t)$:

$$\dot{x}(t) = -i\frac{e}{m}\int\limits_{-\infty}^{\infty}\frac{\omega'\,E_x(\omega')\,\exp(-i\,\omega'\,t)}{\omega_0^2-\omega'^2-2\,i\,\omega'\,\delta_0}\frac{d\omega'}{2\,\pi}. \tag{1.58}$$

Substituting (1.58) and the representation of the field $E_x(t)$ in terms of the Fourier transform (1.13) into (1.57), then averaging over field states with the help of (1.54) and integrating over positive frequencies, we find for the average power

$$P \equiv \langle P\rangle = \frac{2\,\pi\,e^2}{3\,m}\int\limits_0^{\infty}\frac{4\,\omega'^2\,\delta_0\,\rho(\omega')\,d\omega'}{\left(\omega_0^2-\omega'^2\right)^2+(2\,\omega'\,\delta_0)^2} \approx \frac{2\,\pi^2\,e^2}{3\,m}\int\limits_0^{\infty}G^{(h)}(\omega')\,\rho(\omega')\,d\omega', \tag{1.59}$$

where

$$G^{(h)}(\omega') = \frac{(\delta_0/\pi)}{(\omega_0-\omega')^2+(\delta_0)^2} \tag{1.60}$$

is the homogeneous line shape. In the derivation of (1.59), we also took into account the fact that, in a random electromagnetic field, all polarizations are equiprobable, so $\left\langle E_x^2\right\rangle = \left\langle \mathbf{E}^2\right\rangle/3$. The approximate equality in the formula (1.59) corresponds to the assumption of weak damping of the harmonic oscillator $\delta_0 \ll \omega_0$. In the limit of zero damping $\delta_0 \to 0$, the line shape (1.60) is approximated by the delta function:

$$G^{(h)}(\omega') \to \delta(\omega'-\omega_0). \tag{1.61}$$

The expression (1.59) describes the energy that is absorbed by the harmonic oscillator per unit time under the action of radiation with spectral density $\rho(\omega')$. For an electromagnetic field with a broad spectrum, much broader than the width of the harmonic oscillator line $\Delta\omega \gg \delta_0$, the replacement (1.61) can be made on the right-hand side of (1.59). As a result, we obtain

$$P = \frac{2\,\pi^2\,e^2}{3\,m}\,\rho(\omega_0). \tag{1.62}$$

The relation (1.62) describes, in particular, interaction of thermal radiation with the harmonic oscillator when the condition $\Delta\omega \gg \delta_0$ is satisfied. Clearly, the absorbed power is defined by the spectral density of radiant energy at the oscillator eigenfrequency.

In a state of dynamic equilibrium between thermal radiation and the oscillator, the energy of (1.62) absorbed per unit time by the oscillator should be compensated for by energy that is expended per unit time for radiation, as given by (1.47):

$$P = A_{sp}\,\mathrm{E}. \tag{1.63}$$

Using the expressions (1.48) and (1.62), we thus obtain the relation between the spectral density of energy of thermal (blackbody) radiation $\rho_T(\omega)$ and the energy E of an oscillator in a state of thermodynamic equilibrium with this radiation:

$$\rho_T(\omega_0) = \frac{\omega_0^2}{\pi^2\, c^3}\,\mathrm{E}. \tag{1.64}$$

Equation (1.64) was obtained by M. Planck at the end of the nineteenth century. It was used to derive the formula for $\rho_T(\omega)$ which marked the beginning of the new quantum physics.

1.4 Morse Oscillator in an Electromagnetic Field

1.4.1 Definition and Main Characteristics

The harmonic oscillator model describes real physical systems in the limit of small deviation from the equilibrium position. As this deviation grows, more and more anharmonicity begins to show itself. The mathematical manifestation of anharmonicity consists in the fact that the potential energy is no longer described by the quadratic coordinate dependence (1.1). Physically, anharmonicity manifests itself first of all in nonlinear processes that are absent for harmonic systems.

An important model of an anharmonic oscillator is the Morse model, in which the potential energy is given by

$$U^{(Morse)}(x) = D\left\{\exp(-2\,k\,x) - 2\,\exp(-k\,x)\right\}, \tag{1.65}$$

where $\underline{D}$ is the binding energy, k the potential parameter, and x the displacement of the oscillator coordinate from the equilibrium position. The Morse model well describes oscillations of atoms in diatomic molecules. In this case $x = r - r_e$, where r and r_e are the current and equilibrium distances between the nuclei [not to be confused with the classical electron radius (1.49)].

In the limit of small deviations from the equilibrium position $x < 1/k$, the Morse potential turns into the parabolic dependence of the potential energy on the coordinate displacement that characterizes a harmonic oscillator: $U^{(Morse)}(x) \approx D\left\{k^2 x^2 - 1\right\}$. Comparing this approximation with the potential energy of the harmonic oscillator $U^{(harm)}(x) = m\,\omega_0^2 x^2/2 - D$, we find the expression $k = \omega_0\,\sqrt{m/2\,D}$ for the Morse potential parameter in terms of the characteristics of the harmonic oscillator.

Figure 1.6 is a plot of the Morse potential constructed for parameters describing a carbon monoxide (CO) molecule and the corresponding parabolic approximation.

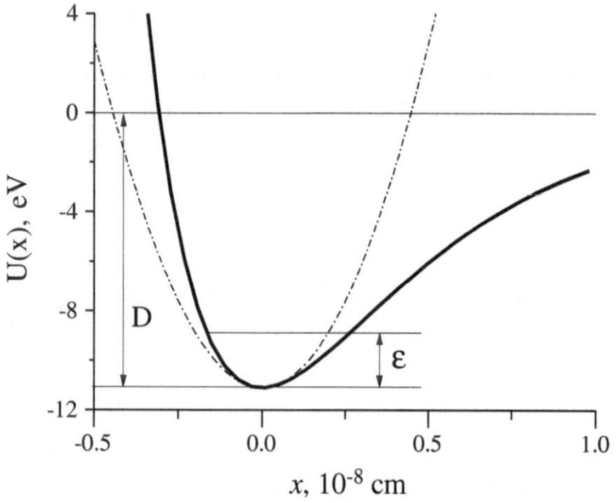

Fig. 1.6 The Morse potential constructed for parameters of a CO molecule (*solid thick curve*) and the corresponding harmonic approximation (*dash-and-dot curve*)

For a CO molecule we have: $D = 11.1$ eV, $\hbar\omega_0 = 0.27$ eV, $m = 1.2 \cdot 10^4\, m_e$ (m_e is the electron mass), $k = 1.19\, a_B^{-1}$ ($a_B \cong 0.53 \cdot 10^{-8}$ is the Bohr radius), $r_e \cong 1.128 \cdot 10^{-8}$ cm.

A qualitative distinction between the Morse potential and the harmonic potential is its asymmetric form with horizontal asymptote $U = 0$ in the region $x \gg 1/k$. This asymptote divides the energy spectrum of the oscillator into two parts: the positive part $(U > 0)$ and the negative part $(U < 0)$. The negative part of the spectrum corresponds to finite motion, when the oscillator coordinate varies between finite limits. In contrast, the positive part of the spectrum corresponds to infinite motion, when the oscillator coordinate increases indefinitely (oscillations turn into recession). When the Morse potential describes a diatomic molecule, in the region of negative energies, the atoms forming the molecule oscillate in a bounded region of space. The positive part of the spectrum of a diatomic molecule corresponds to a dissociated state, in which the interatomic distance tends to infinity. In this context, the binding energy D has the meaning of a molecule dissociation energy. Thus the Morse potential describes not only oscillatory motion, but also dissociation of atoms.

Within the framework of the quantum–mechanical formalism, the region of negative energies corresponds to a discrete spectrum, where the system energy changes stepwise (in a quantum manner). Positive energies correspond to a continuous spectrum, in which the system energy can change by an arbitrarily small value. The harmonic oscillator in the quantum picture has only a discrete spectrum, in which the energy levels are evenly spaced. The Morse oscillator has both a discrete and a continuous energy spectrum, and the energy levels in the discrete spectrum are not evenly spaced: the distance between them decreases as one approaches the edge of the continuous spectrum.

In classical language, nonequidistance of levels corresponds to anharmonicity of oscillations of the oscillator, and this anharmonicity is characterized by the dimensionless parameter x_e, with value 0.00612 for the CO molecule. The anharmonicity parameter serves as a measure of the deviation of the discrete spectrum of a real oscillator from the equidistant approximation. In the Morse model the anharmonicity parameter is unequivocally related to other characteristics of the potential according to the formula $x_e = \hbar\,\omega_0/4\,D$ Hence for a CO molecule we obtain the theoretical value for the anharmonicity parameter $x_e = 0.00608$. Comparing this value with the above experimental value, we come to the conclusion that in this case the Morse model gives a relative error of only 0.65 %. Thus the Morse potential is a very good approximation to a real potential in the case of diatomic molecules.

1.4.2 Free Oscillations of the Morse Oscillator

In the classical context, nonequidistance of a discrete energy spectrum corresponds to the fact that the period of oscillations of the Morse oscillator depends on its energy, in contrast to the constant period of oscillations of the harmonic oscillator $T = 2\,\pi/\omega_0$. To determine this dependence and other peculiarities of the motion, we begin by considering free oscillations of the Morse oscillator in the negative part of the spectrum, when $\varepsilon < D$ [ε is the oscillator energy measured from the bottom of the potential well (Fig. 1.6)]. It is convenient to introduce the dimensionless displacement $\rho = k\,x$ of the coordinate from the equilibrium position and the dimensionless time $\tau = \omega_0\,t$. Then the equation of motion of the Morse oscillator can be written in the form

$$\ddot{\rho}_{\tau^2} = \exp(-2\,\rho) - \exp(-\rho). \tag{1.66}$$

It should be noted that the dimensionless equation (1.66) does not contain the binding energy *D* and the potential parameter *k*, that is, it is universal for the given kind of potential energy.

To determine the dependence $\rho(\tau)$, it is better not to solve the equation of motion (1.66) directly, but to use the energy conservation law which becomes, in dimensionless variables,

$$\ddot{\rho}_\tau^2 + e^{-2\,\rho} - 2\,e^{-\rho} = \tilde{\varepsilon} - 1, \tag{1.67}$$

where $\tilde{\varepsilon} = \varepsilon/D$ is the dimensionless energy. For finite motion (oscillation mode) $1 > \tilde{\varepsilon} > 0$. From (1.67) it is easy to obtain

$$\tau - \tau_0 = \int\limits_{\rho_0}^{\rho} \frac{dy}{\sqrt{\tilde{\varepsilon} - 1 + 2\,e^{-y} - e^{-2\,y}}}, \tag{1.68}$$

where τ_0 is the integration constant. The formula (1.68) gives the dependence $\rho(\tau)$ in implicit form. To obtain the explicit form of the function $\rho(\tau)$, one must calculate the integral on the right-hand side of (1.68) and express ρ in terms of the dimensionless time τ. As a result, we find

$$\rho(\tau) = \ln\left\{\frac{1 - \sqrt{\tilde{\varepsilon}}\,\sin\left[\sqrt{1-\tilde{\varepsilon}}\,\tau + \chi(\tilde{\varepsilon}, \rho_0)\right]}{1-\tilde{\varepsilon}}\right\}, \tag{1.69}$$

where $\chi(\tilde{\varepsilon}, \rho_0)$ is the initial phase of oscillations defined by the initial condition $\rho_0 = \rho(\tau = 0)$. The resulting expression is valid in the dimensionless energy range $1 > \tilde{\varepsilon} > 0$, that is, in the negative part of the spectrum.

The harmonic approximation is true for low energies of excitation of the Morse oscillator, that is, for $\tilde{\varepsilon} \ll 1$ $(\varepsilon \ll D)$. Then (1.69) turns into the well-known expression for free oscillations of the harmonic oscillator (in dimensionless variables):

$$\rho(\tau, \tilde{\varepsilon} \ll 1) \cong -\sqrt{\tilde{\varepsilon}}\sin(\tau + \chi). \tag{1.70}$$

As expected, the oscillation amplitude is proportional to the square root of the energy, and the oscillations themselves follow the harmonic law.

From (1.69) it follows that the period of oscillations of the Morse oscillator is energy-dependent and equal to

$$T^{(Morse)}(\varepsilon) = \frac{2\pi}{\omega_0\sqrt{1 - \varepsilon/D}}. \tag{1.71}$$

Hence the period grows with increasing energy and as $\varepsilon \to D$, one finds $T^{(Morse)} \to \infty$. Thus, at the boundary of the negative part of the spectrum ($\tilde{\varepsilon} = 1$), the periodic motion of the Morse oscillator transforms into aperiodic motion. In dimensionless variables, the law of motion for the boundary energy $\varepsilon = D$ is

$$\rho(\tau) = \ln\left\{\frac{1}{2}\left[1 + \left(\tau - \sqrt{2\exp(\rho_0) - 1}\right)^2\right]\right\}, \tag{1.72}$$

where $\rho_0 = \rho(0)$ is the value of the dimensionless coordinate at the initial time, and one must have $\rho_0 \geq -\ln 2$ for the radicand on the right-hand side of (1.72) to be positive. In the limit of long times $\tau \to \infty$, the dimensionless coordinate of the Morse oscillator at the spectrum edge grows logarithmically: $\rho \propto \ln \tau$. In this case, the dimensionless velocity is obviously $\dot{\rho}_\tau \to 1/\tau$, and hence decreases to zero at infinity.

For energies in the positive part of the spectrum $\tilde{\varepsilon} > 1$, calculation of the integral (1.68) results in the law of motion

$$\rho(\tau) = \ln\left\{\frac{\tilde{v}^2 + [A(\tilde{v},\rho_0)\exp(-\tilde{v}\,\tau) - 1]^2}{2\tilde{v}^2 A(\tilde{v},\rho_0)\exp(-\tilde{v}\,\tau)}\right\}, \tag{1.73}$$

where

$$A(\tilde{v},\rho_0) = \tilde{v}^2 \exp(\rho_0) + 1 + \tilde{v} \exp(\rho_0)\sqrt{\tilde{v}^2 + 2\exp(-\rho_0) - \exp(-2\rho_0)} \quad (1.74)$$

is the constant defined by the initial conditions and $\tilde{v} = \sqrt{\tilde{\varepsilon} - 1}$ is the dimensionless velocity. From (1.73), it follows that, for long times $\tau \gg \tilde{v}^{-1}$, the oscillator coordinate is $\rho \propto \tilde{v}\tau$. So for a positive total energy of the Morse oscillator ($\tilde{\varepsilon} > 1$), linear recession occurs.

The time dependences of free motion of the Morse oscillator in the three above modes are presented in Fig. 1.7 for zero initial coordinate ($\rho_0 = 0$).

Figure 1.7 shows that, for increasing total energy, oscillations (curve 1) change into recession under the logarithmic law (curve 2) for zero total energy, and finally linear recession (in the time interval $\tau \gg 1$) for positive energy (curve 3). Oscillations of the Morse oscillator, in contrast to the harmonic oscillator, are of an asymmetric nature. This is connected with the asymmetry of the Morse potential (Fig. 1.6).

The three modes of one-dimensional motion of the Morse oscillator described here have an analogy in the two-dimensional case. Motion in an attractive Coulomb field or a gravitational field can be elliptic (negative energy), parabolic (zero energy), or hyperbolic (positive energy).

1.4.3 Morse Oscillator in an Electromagnetic Radiation Field

Let the charged Morse oscillator be acted on by a Gaussian electric field pulse (1.27). The equation of motion of the oscillator in dimensionless variables can be written in the form [3]

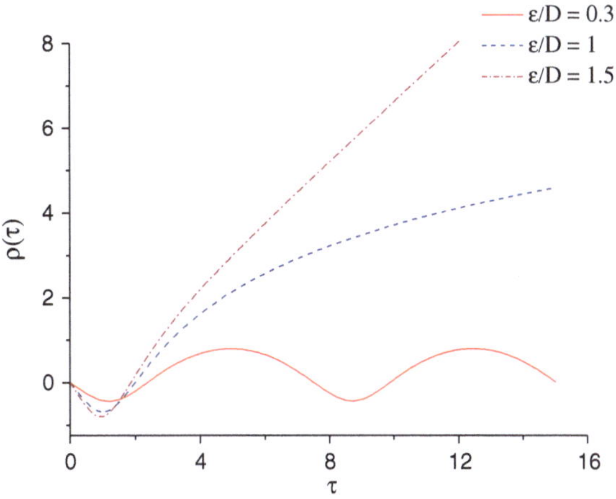

Fig. 1.7 Three modes of free motion of the Morse oscillator: 1—oscillation mode $\tilde{\varepsilon} = 0.3$. 2—recession in the mode of zero total energy. 3—recession in the positive energy mode $\tilde{\varepsilon} = 1.5$

$$\ddot{\rho}_{\tau^2} = \exp(-2\,\rho) - \exp(-\rho) + \gamma\,\tilde{E}(\tau), \tag{1.75}$$

where $\gamma = k\,q\,E_0 / m\,\omega_0^2$ is the dimensionless parameter characterizing the binding force between the electric field and the Morse oscillator with q the effective charge of the oscillator, and $\tilde{E}(\tau, \varphi_0) = E(\tau/\omega_0)/E_0$ is the dimensionless electric field strength as a function of the dimensionless time $\tau = \omega_0\,t$.

The effective charge of the oscillator in the case of a diatomic molecule can be determined by the formula $q = \partial\mu(r)/\partial r$, where $\mu(r)$ is the dependence of the dipole moment of the diatomic molecule on the distance between the nuclei (not to be confused with the constant dipole moment of the molecule, if any). The effective charge of the oscillator describing a CO molecule, calculated using the "Gaussian" quantum-chemical program, is $q = 0.78\,e$ (here e is the elementary charge). For the interaction between an electric field and a carbon monoxide molecule, the above numerical values for characteristics of the Morse potential can be used to obtain the following expression for the binding force parameter as a function of the amplitude of the electric field strength and the atomic strength: $\gamma = 0.65\,E_0/E_a$, where $E_a \cong 5 \cdot 10^9$ V/cm. Thus, the parameter γ is close to unity for rather high values of the electric field strength (of the order of the atomic strength).

Numerical analysis shows that solution of (1.75) differs from its harmonic analog for $\gamma > \gamma^* = 0.05$. The critical value γ^* depends weakly on the carrier frequency, initial phase, and pulse width in the ultrashort pulse mode $\eta < 10$.

Thus in the region of weak binding force $\gamma < 0.05$, the phase dependence of excitation of the anharmonic oscillator in the Morse model is described by the same formulas as for the harmonic oscillator. Using the given inequation for the pulse width $\Delta t < 30$ fs, we obtain the following restriction on the energy flux density in an exciting pulse: $d\mathrm{E}/dS < 10$ J/cm^2 When this holds, the harmonic approximation for a CO molecule is valid.

Initially at rest, the Morse oscillator will gain energy according to the formula

$$\varepsilon = 2\,\gamma D \int_{-\infty}^{\infty} \dot{\rho}(\tau)\,\tilde{E}(\tau)\,d\tau \tag{1.76}$$

under the action of the electric field. This follows from the standard expression (1.57) for the power. Here, as before, energy is measured from the bottom of the potential well (Fig. 1.6). The calculated dependence of the Morse oscillator energy on the parameter γ after the action of a single-cycle Gaussian electric field pulse (1.27) with constant CE phase is presented in Fig. 1.8 for different values of the CE phase.

We thus find that the dependence of the Morse oscillator energy on the field force for an ultrashort pulse depends essentially on the initial (CE) phase. As the phase varies from $-\pi/2$ to $\pi/2$, this dependence changes from a monotonically increasing curve to a curve with a minimum at $\gamma = 0.52$. In this case it thus follows that diatomic molecule dissociation (that is, reaching the value $\tilde{\varepsilon} = 1$) proceeds

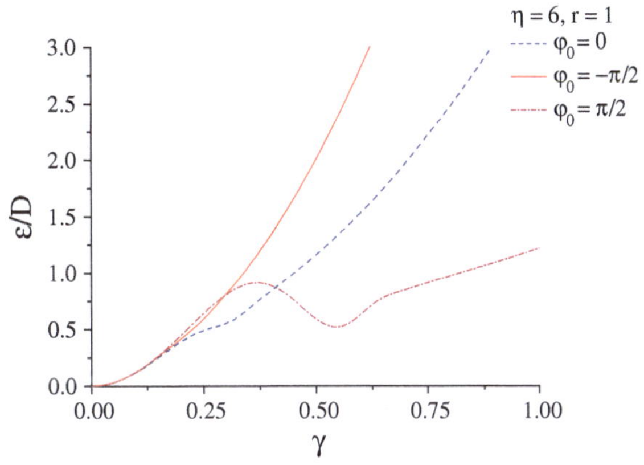

Fig. 1.8 Normalized energy of the Morse oscillator after the action of a single-cycle electric pulse ($\eta = \omega_0\,\Delta t$) as a function of the binding force for different values of the CE phase

more easily for the CE phase $\varphi_0 = -\pi/2$. But if $\varphi_0 = \pi/2$, dissociation by a pulse $\eta = 4$ is possible only for $\gamma > 1$.

1.4.4 Morse Oscillator Radiation

According to the general laws of electrodynamics, the excited charged Morse oscillator will radiate electromagnetic waves. As in the case of the harmonic oscillator, the Morse oscillator radiation power at times $t > \Delta t$ can be expressed in terms of the oscillator energy ε (see Appendix II):

$$Q^{(Morse)} = \frac{2q^2\,\omega_0^2}{3\,m\,c^3}\,\varepsilon\sqrt{1-\frac{\varepsilon}{D}}. \tag{1.77}$$

Comparing the expression (1.77) with the formulas (1.47)–(1.48) describing the harmonic oscillator, we come to the conclusion that the Morse oscillator dipole radiation power as a function of the energy ε differs from its harmonic analog by the multiplier $\sqrt{1-\varepsilon/D}$. Thus for excitation of the charged Morse oscillator in the range of energy values near the dissociation energy $\varepsilon \approx D$, the radiation power tends to zero. This is connected with the decrease in the oscillation frequency for increasing oscillator energy. From (1.77), it follows that the maximum dipole radiation power of the Morse oscillator is achieved at the energy $\varepsilon = 2D/3$. The linear mode of radiation excitation takes place for low energies $\varepsilon \ll D$. An appreciable deviation from linearity (more than 10 %) arises when $\varepsilon > D/5$

Figure 1.9 shows the results of calculation of the Morse oscillator dipole radiation power as a function of the CE phase after the action of a subcycle electric field pulse for different values of the field force parameter γ and $\omega = \omega_0$ ($r = 1$).

From Fig. 1.9 it follows that sensitivity of the Morse oscillator radiation power to the CE phase after excitation by a single-cycle pulse is observed only in the nonlinear mode $\gamma > 0.1$. Otherwise ($\gamma < 0.1$), when the harmonic approximation

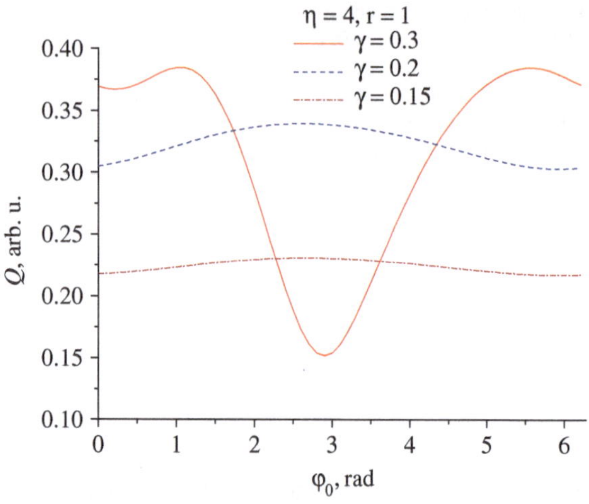

Fig. 1.9 Phase dependence of the charged Morse oscillator dipole radiation power after the action of a subcycle Gaussian electric field pulse (1.27) for different binding force parameters and $\omega = \omega_0$

is justified, the phase dependence disappears. For the harmonic oscillator, the expression (1.38) for the radiation power phase modulation factor is valid. Substituting in the numerical values of the parameters, we find $K(r = 1,\ \eta = 4) \approx 2 \cdot 10^{-7}$.

Thus for single-cycle and longer exciting pulses with controlled CE phase, the phase dependence of the Morse oscillator radiation power manifests itself only in the nonlinear excitation mode. Then the behavior of the phase modulation differs from the harmonic law (1.51) and strongly depends on the value of the parameter γ describing the binding between the field and the oscillator. Appreciable phase sensitivity of the Morse oscillator excitation in the linear mode arises only for subcycle field pulses at $\eta < 2$ ($\Delta t < 3$), when the formula (1.38) is valid.

1.4.5 Morse Oscillator in a Chirped Pulse Field

As in the case of the harmonic oscillator, excitation of the Morse oscillator by a chirped pulse exhibits a stronger dependence on the phase parameter (frequency chirp). Oscillations of the Morse oscillator calculated by numerical solution of (1.75) for different values of the dimensionless chirp ($\alpha = \kappa\,\Delta t^2$) are shown in Fig. 1.10.

In this case, it can be seen that the oscillation amplitude varies considerably in going from one value of the parameter $\alpha = \kappa\,\Delta t^2$ to another. In particular, the minimum amplitude corresponds to the value $\alpha = 0.68$.

From the formula (1.69), it follows that the amplitude of oscillations of the Morse oscillator is defined by the oscillator energy according to the expression

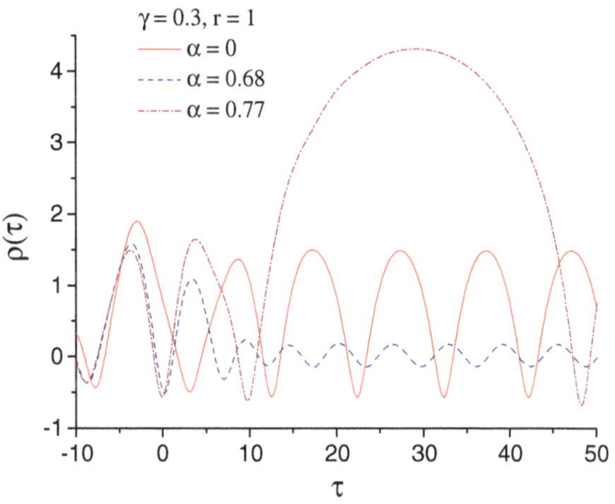

Fig. 1.10 Oscillations of the Morse oscillator excited by two-cycle pulses ($\eta = 12$) with different values of the frequency chirp and with interaction constant $\gamma = 0.3$: *Solid curve* $\alpha = 0$. *Dash-and-dot curve* $\alpha = 0.68$. *Dashed curve* $\alpha = 0.77$

$$\rho_{amp}(\tilde{\varepsilon}) = \frac{1}{2}\ln\left\{\frac{1+\sqrt{\tilde{\varepsilon}}}{1-\sqrt{\tilde{\varepsilon}}}\right\}, \tag{1.78}$$

so it is clear that the greater the excitation energy, the greater the oscillation amplitude.

To explain the dependences of Fig. 1.10, Fig. 1.11 shows the dimensionless energy of the Morse oscillator as a chirp function for single-cycle and two-cycle exciting pulses. $\tilde{\varepsilon}(\alpha)$ was calculated using (1.76) with numerical solution of (1.75).

Here we see that, with growing pulse width, the dependence of the excitation energy on the chirp value becomes stronger, and the function $\tilde{\varepsilon}(\alpha)$ becomes more asymmetric in relation to the change of chirp sign. For a two-cycle pulse at the specified values of the parameters, the excitation energy has a deep minimum for $\alpha = 0.68$. This is reflected in the corresponding time dependence of Fig. 1.10 by a small value of the asymptotic oscillation amplitude. In contrast, Fig. 1.11 shows that, for $\alpha = 0.77$, the Morse oscillator excitation energy has a local maximum, so according to the formula (1.78) oscillations corresponding to this chirp value have a large amplitude. The straight line in the same figure shows the value $\tilde{\varepsilon}_{opt} = 2/3$ of the excitation energy which maximizes the Morse oscillator radiation power after cessation of action of the pulse.

Thus using the Morse potential in the framework of the classical approach allows one to simulate the interaction of electromagnetic radiation with molecular systems in order to investigate their excitation and dissociation.

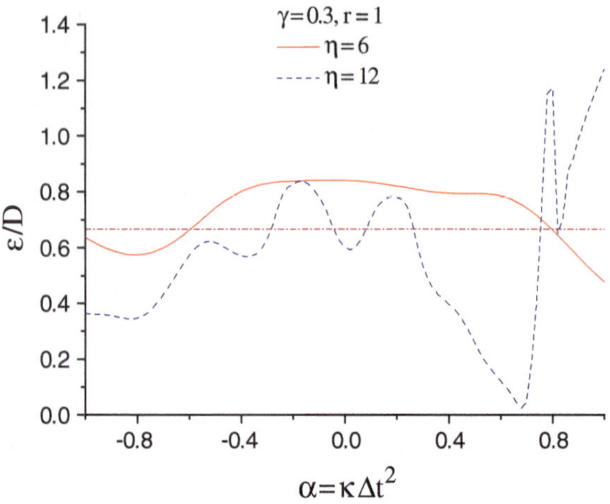

Fig. 1.11 Dimensionless energy of the Morse oscillator as a chirp function for single-cycle and two-cycle exciting pulses ($\gamma = 0.3$, $r = 1$). The *straight line* shows the optimum energy value, at which the oscillator radiation power, averaged over the period after cessation of action of the electromagnetic pulse, is maximum

Appendix I

Here we obtain the relation (1.53) between the spectral density of energy of a random stationary field and the autocorrelation function of the electric field strength. By definition, the spatial density of the energy E of the uniform electromagnetic field occupying the volume V is

$$u = \frac{\mathrm{E}}{V} = \left\langle \frac{\mathbf{E}^2 + \mathbf{H}^2}{8\,\pi} \right\rangle = \frac{\langle \mathbf{E}^2(t) \rangle}{4\,\pi}. \tag{A.1}$$

The angle brackets denote averaging over the electromagnetic field state, that is, over amplitudes, phases, and polarizations of its monochromatic components. In the third equality of (A.1), we use the fact that, for an electromagnetic wave in vacuum, the strengths of the electric and magnetic fields coincide. By definition, the stationary electric field strength correlator is

$$KE_{ik}(\tau) = \langle E_i(t)\, E_k(t+\tau) \rangle = \int\limits_{-\infty}^{\infty} KE_{ik}(\omega)\, \exp(-i\,\omega\,\tau)\, \frac{d\omega}{2\,\pi}. \tag{A.2}$$

The Fourier transform of the correlator $KE_{ik}(\omega)$ is called the field *spectral density tensor*.

Using (A.2), (A.1) can be rewritten as

$$u = \frac{\langle \mathbf{E}^2(t) \rangle}{4\,\pi} = \frac{KE_{ii}(\tau = 0)}{4\,\pi} = \frac{1}{(2\,\pi)^2} \int\limits_0^{\infty} KE_{ii}(\omega)\, d\omega. \tag{A.3}$$

Here the Einstein summation convention is implied. In going to integration over positive frequencies alone, we use the fact that the function $KE_{ii}(\omega)$ is even, which is easily shown from its definition and from the reality of the electric field strength $E_i(t)$.

We now compare (A.3) with the determination of the spectral density of the field

$$u = \int_0^\infty \rho(\omega)\, d\omega, \tag{A.4}$$

from which it follows that

$$\rho(\omega) = \frac{1}{(2\pi)^2} KE_{ii}(\omega), \tag{A.5}$$

$$KE_{ii}(\omega) = \int_{-\infty}^{\infty} KE_{ii}(\tau)\, e^{i\omega\tau}\, d\tau = \int_{-\infty}^{\infty} e^{i\omega\tau} \langle \mathbf{E}(t)\,\mathbf{E}(t+\tau)\rangle\, d\tau. \tag{A.6}$$

Equations (A.5)–(A.6) imply (1.53).

Appendix II

Here we derive the formula (1.77) for the power of dipole radiation of the charged Morse oscillator averaged over the oscillation period after cessation of an exciting electromagnetic field pulse ($t \gg \Delta t$), when the oscillations can be considered to be free. We proceed from the expression for the instantaneous dipole radiation power of one-dimensional oscillations of the oscillator with energy ε:

$$Q(t, \varepsilon) = \frac{2\left|\ddot{d}(t, \varepsilon)\right|^2}{3\, c^3}, \tag{A.7}$$

where $d(t, \varepsilon) = q\, x(t, \varepsilon)$ is the dipole moment of the oscillator, $k = \omega_0 \sqrt{m/2D}$ is the Morse potential parameter, D is the binding energy, and m is the oscillator mass. Let us rewrite (A.1) in dimensionless variables $\rho = kx$ and $\tau = \omega_0 t$ and average over the dimensionless oscillation period

$$\tilde{T}^{(Morse)} = T^{(Morse)} \omega_0 = \frac{2\pi}{\sqrt{1-\tilde{\varepsilon}}}, \tag{A.8}$$

where $\tilde{\varepsilon} = \varepsilon/D$ is the normalized energy. The result is

$$\langle Q(t, \varepsilon)\rangle_T = \frac{4\, q^2\, \omega_0^2\, D}{3\, m\, c^3} \langle \ddot{\rho}_{\tau^2}^2 \rangle_{\tilde{T}}. \tag{A.9}$$

We must therefore calculate the average of the squared dimensionless second derivative $\langle \ddot{\rho}^2_{\tau^2} \rangle_{\tilde{T}}$. This average can be written

$$\langle \ddot{\rho}^2_{\tau^2} \rangle_{\tilde{T}} = \frac{2}{\tilde{T}} \int\limits_{\rho_1(\tilde{\varepsilon})}^{\rho_2(\tilde{\varepsilon})} \frac{\ddot{\rho}^2}{\dot{\rho}} \, d\rho = 2 \frac{\sqrt{1-\tilde{\varepsilon}}}{2\pi} I_M(\tilde{\varepsilon})$$

$$= 2 \frac{\sqrt{1-\tilde{\varepsilon}}}{2\pi} \int\limits_{\rho_1(\tilde{\varepsilon})}^{\rho_2(\tilde{\varepsilon})} \frac{(\exp(-2\rho) - \exp(-\rho))^2}{\sqrt{\tilde{\varepsilon} - 1 + 2\exp(-\rho) - \exp(-2\rho)}} d\rho, \quad \text{(A.10)}$$

where $\rho_1 = -\ln\left(1 + \sqrt{\tilde{\varepsilon}}\right)$ and $\rho_2 = -\ln\left(1 - \sqrt{\tilde{\varepsilon}}\right)$ are the turning points at which the oscillator velocity is zero. They correspond to zeros of the radicand in the denominator of the right-hand side of (A.10). In writing (A.10), we used the fact that the electromagnetic pulse had ceased and that the dynamics of the Morse oscillator is defined solely by its potential energy, validating the expression for acceleration of the free oscillator (1.75) and its velocity following from (1.67).

It is not difficult to calculate the integral $I_M(\tilde{\varepsilon})$ on the right-hand side of (A.10) if we make the change of variable $\rho = -\ln\left(1 + \sqrt{\tilde{\varepsilon}} y\right)$. Then after elementary transformations we find

$$I_M(\tilde{\varepsilon}) = \tilde{\varepsilon} \int\limits_{-1}^{1} \frac{y^2}{\sqrt{1-y^2}} dy = \frac{\pi}{2} \tilde{\varepsilon}. \quad \text{(A.11)}$$

Substituting the right-hand side of (A.6) into (A.5) and the resulting expression into (A.4), we arrive at the formula (1.77) $\tilde{\varepsilon} = \varepsilon / D$:

$$\langle Q(t, \varepsilon) \rangle_T = \frac{2 q^2 \omega_0^2 D}{3 m c^3} \tilde{\varepsilon} \sqrt{1 - \tilde{\varepsilon}}. \quad \text{(A.12)}$$

References

1. Kundu, M., Kaw, P.K., Bauer, D.: Phys. Rev. A **85**, 023202 (2012)
2. Arustamyan, M.G., Astapenko, V.A.: Laser Physics **18**, 1031 (2008)
3. Astapenko, V.A., Romadanovskii, M.S.: JETP **110**, 376 (2010)

Chapter 2
Interaction of Ultrashort Electromagnetic Pulses with Matter: Description in the Framework of Perturbation Theory

Considerable advances have been made in the generation of ultrashort electromagnetic field pulses of controlled shape over a wide spectral range [1]. In the infrared, visible, and far-ultraviolet spectral regions, pulses have been produced with widths equal to the period of the electromagnetic field oscillation at the carrier frequency (single-cycle pulses). Single-cycle pulses have widths of a few femtoseconds in the near–infrared and visible regions and of the order of a hundred attoseconds and less in the far-UV range. Such pulses provide the basis for studying electron dynamics with resolutions approaching one atomic time unit (24 as). In the terahertz range, half-cycle pulses are generated with widths of the order of a picosecond, which is promising in particular for quantum calculations using Rydberg states [2].

With the development of techniques for generating ultrashort pulses of electromagnetic radiation, it has become urgent to consider the peculiar features of interactions between radiation and specific substances, and to develop adequate ways of describing electromagnetic processes in ultrashort fields. The use of ultrashort laser pulses opens up new possibilities in superfast monitoring of light-induced phenomena, attosecond metrology, spectroscopy, microscopy, and plasmonics.

The interaction between single-cycle and subcycle pulses and matter has characteristic features that differ from the case of multicycle pulses. For the latter, it is well known that the photoprocess probability does not depend on the phase of the electromagnetic field, but is determined by the strength and carrier frequency. An important peculiarity of ultrashort interactions is the dependence of the photoprocess probability on the phase characteristics of the radiation, and in particular the carrier phase with respect to the envelope and the frequency chirp. (The frequency chirp in the linear case is a coefficient determining the time dependence of the frequency.) This phase dependence can be used both to obtain information about wave functions of atomic electrons and also to use the phase method to control light-induced processes.

In the previous chapter we considered the elementary case where the interaction of a substance with ultrashort pulses can be described by a classical oscillator. Hereafter (Chaps. 2 and 3), these interactions will be investigated in the framework

V. Astapenko, *Interaction of Ultrashort Electromagnetic Pulses with Matter*, SpringerBriefs in Physics, DOI: 10.1007/978-3-642-35969-9_2,

of a more realistic approach, the semiclassical theory, where the substance is described at the quantum level and the radiation is described classically.

In this chapter it is assumed that the electric field in the ultrashort pulse is not too high (or the photoprocess cross-section is small enough) to justify applying the quantum–mechanical perturbation theory.

2.1 Derivation of the Basic Formula

When perturbation theory is applicable and one considers the action of relatively long radiation pulses on a substance, the response to the electromagnetic action is usually described with reference to the photoprocess probability per unit time w.

This can be determined from the cross-section σ of the photoprocess by the formula

$$w = \int \sigma(\omega') \frac{I(\omega')}{\hbar\,\omega'}\, d\omega', \tag{2.1}$$

where $I(\omega')$ is the spectral intensity of radiation which, for a monochromatic field of frequency ω, is $I(\omega') = I_0\,\delta(\omega - \omega')$, with I_0 the integrated intensity.

It should be noted that the cross-section $\sigma(\omega')$ corresponds to the action of a *monochromatic* field at a specified frequency ω', while the integration over frequencies on the right-hand side of (2.1) takes into account the non-monochromaticity of the radiation. If the pulse lasts much longer than the period of oscillation $T = 2\pi/\omega$ at the carrier frequency, the spectral intensity of radiation is determined in terms of the electric field strength autocorrelator by the formula

$$I(\omega') = \frac{c}{(2\,\pi)^2} \int\limits_{-\infty}^{\infty} \langle E_i(t)\,E_i(t+\tau)\rangle_t \exp(i\,\omega'\,\tau)\,d\tau, \tag{2.2}$$

where the symbol $\langle\ldots\rangle_t$ denotes the time average (see the derivation of (2.2) in Appendix 1 and note that $I(\omega') = c\,\rho(\omega')$).

In going to ultrashort pulses consisting of a small number of optical cycles or even of part of a cycle, the concept of photoprocess probability per unit time becomes inadequate, and a description of the electromagnetic interaction through the total probability for the whole time of action of the pulse is more physically meaningful. So for ultrashort pulses, the formulas (2.1)–(2.2) are no longer valid.

To describe processes in ultrashort fields within the framework of the perturbation theory, (2.1) is replaced by a calculation of the *total* photoprocess probability W *for the whole time of action* of the ultrashort pulse, retaining the description of the properties of the excited system in terms of the cross-section $\sigma(\omega')$. This is the problem to which this section is dedicated.

So let us consider photoexcitation of a quantum system from the ground state $|0\rangle$ to some excited state $|n\rangle$ under the action of a dipole perturbation

$$\hat{V}(t) = -\hat{d}_i E_i(t), \tag{2.3}$$

where $\hat{\mathbf{d}}$ is the electric dipole moment operator of the system and $\mathbf{E}(t)$ is the electric field strength, taken to be a classical quantity that is independent of the spatial coordinate (dipole approximation). To first order in perturbation theory, the amplitude of this process for the whole time of action of the field is

$$c_{n0} = -\frac{i}{\hbar} \int_{-\infty}^{\infty} \langle n|\hat{d}_i(t)|0\rangle\, E_i(t)\, dt, \tag{2.4}$$

where $\hat{d}_i(t) = \exp\left(i\hat{H}_0 t/\hbar\right) \hat{d}_i \exp\left(-i\hat{H}_0 t/\hbar\right)$ is the dipole moment operator in the interaction representation and $\hat{H}_0$ is the zero-order Hamiltonian of the system.

The probability of photoexcitation with the transition $|0\rangle \rightarrow |n\rangle$ for the whole time of action of the perturbation is

$$W_{n0} = \frac{1}{\hbar^2} \int_{-\infty}^{\infty} \int_{-\infty}^{\infty} \langle 0|\hat{d}_i(t')|n\rangle \langle n|\hat{d}_k(t)|0\rangle E_i(t')\, E_k(t)\, dt\, dt'. \tag{2.5}$$

The total photoexcitation probability taking into account transitions to all states of the system is then

$$W_{tot} = \sum_n W_{n0} = \frac{1}{\hbar^2} \int_{-\infty}^{\infty} \int_{-\infty}^{\infty} \langle 0|\hat{d}_i(t')\, \hat{d}_k(t)|0\rangle E_i(t')\, E_k(t)\, dt\, dt'. \tag{2.6}$$

In the derivation of (2.6) from (2.5), we used the completeness of the set of functions $|n\rangle$, and the fact that $\langle 0|\hat{d}_i(t)|0\rangle = 0$ for a spherically symmetrical system, so to satisfy the completeness condition in the sum (2.6), a summand can be included that corresponds to the invariable state of the system W_{00}.

We now use the fact that the correlator of dipole moments of a system that is stationary in the unperturbed state depends only on the time difference $\tau = t' - t : K_{ik}(\tau) = \langle 0|\, \hat{d}_i(t')\, \hat{d}_k(t)|0\rangle$, something that is easy to check directly. Then the formula (2.6) can be rewritten as

$$W_{tot} = \frac{1}{\hbar^2} \int_{-\infty}^{\infty} \int_{-\infty}^{\infty} K(\tau) E_i(t + \tau)\, E_i(t)\, dt\, d\tau. \tag{2.7}$$

Here we note that, for a centrosymmetric system, $K_{ik}(\tau) = \delta_{ik} K(\tau)$, where $K(\tau) = K_{ii}(\tau)/3$. Replacing the terms in the integrand of (2.7) by Fourier transforms $f(t) = \int_{-\infty}^{\infty} f(\omega') \exp(i\omega' t) \frac{d\omega'}{2\pi}$, we obtain

$$W_{tot} = \frac{2\pi}{\hbar^2 c} \int_{-\infty}^{\infty} \int_{-\infty}^{\infty} K(\omega') I(\omega', t)\, d\omega' dt, \tag{2.8}$$

where we have introduced the function

$$I(\omega', t) = \frac{c}{(2\pi)^2} \int_{-\infty}^{\infty} E_i(t)\, E_i(t+\tau)\, \exp(i\omega'\tau)\, d\tau. \tag{2.9}$$

This function can be called the spectral density of instantaneous radiation intensity. The Fourier transform of the dipole moment correlator is by definition

$$K(\omega) = \frac{1}{3} \int_{-\infty}^{\infty} \langle 0|\, \hat{d}_i(t)\, \hat{d}_i(t+\tau) |0\rangle \exp(i\omega\tau)\, d\tau. \tag{2.10}$$

Starting from (2.9), it is not difficult to prove that

$$\int_{-\infty}^{\infty} I(\omega', t)\, dt = \frac{c}{(2\pi)^2} |\mathbf{E}(\omega')|^2. \tag{2.11}$$

Substituting the expression (2.11) into the right-hand side of (2.8), we obtain

$$W_{tot} = \frac{1}{2\pi\hbar^2} \int_{-\infty}^{\infty} K(\omega')\, |\mathbf{E}(\omega')|^2\, d\omega'. \tag{2.12}$$

Let us apply this to a monochromatic field $\mathbf{E}(t) = \mathbf{E}_0 \cos(\omega t)$, where the average intensity over the period $T = 2\pi/\omega$ is

$$\langle I(\omega', t)\rangle_T = I_0\, \delta(\omega - \omega') \langle 1 + \exp(-2i\omega t)\rangle_T = I_0\, \delta(\omega - \omega'), \tag{2.13}$$

with $I_0 = c\,\mathbf{E}_0^2/8\pi$. Integrating (2.13) over the finite time interval Δt as in (2.8), we find

$$W_{tot} = \frac{2\pi}{\hbar^2 c} K(\omega)\, I_0\, \Delta t, \tag{2.14}$$

whence the photoexcitation probability *per unit time* is

$$w_{tot} = \frac{2\pi}{\hbar^2 c} K(\omega)\, I_0. \tag{2.15}$$

On the other hand, by definition of the photoabsorption cross-section in a monochromatic field of frequency ω, we have

$$w_{tot} = \sigma(\omega) \frac{I_0}{\hbar\omega}. \tag{2.16}$$

The value $I_0/\hbar\,\omega$ represents the monochromatic radiation photon flux density. Comparing (2.15) and (2.16), we find the expression for the Fourier transform of the *dipole moment correlator* in terms of the *photoabsorption cross-section* in the dipole approximation:

$$K(\omega) = \frac{\hbar\,c}{2\,\pi\,\omega}\,\sigma(\omega). \tag{2.17}$$

This implies the important formula for the photoabsorption cross-section:

$$\sigma(\omega) = \frac{2\,\pi\,\omega}{\hbar\,c} K(\omega). \tag{2.18}$$

Substituting (2.17) in (2.12), we obtain the expression for the total probability of the process under consideration for the whole time of action of the radiation pulse. Since the case in point is photoabsorption, the integration in (2.12) is necessarily restricted to positive frequencies, so finally we find

$$W_{tot} = \frac{c}{(2\,\pi)^2}\int_0^\infty \sigma(\omega')\,\frac{|\mathbf{E}(\omega')|^2}{\hbar\,\omega'}\,d\omega'. \tag{2.19}$$

Obtained within the framework of the perturbation theory, the expression (2.19) is the *basic formula* describing the total photoprocess probability under the action of ultrashort pulses. Of course, it is true whenever $W_{tot} < 1$, according to the mathematical definition of probability.

For the photoabsorption cross-section the equation can be written

$$\sigma(\omega') = \frac{2\,\pi^2\,e^2}{m\,c}\left\{\sum_{n>0} f_{n0}\,G_{n0}(\omega') + g_{c0}(\omega')\right\} = \sum_n \sigma_{n0}(\omega') + \sigma_{c0}(\omega'), \tag{2.20}$$

where the contributions to the total cross-section of the discrete spectrum (the sum over n) and the continuous spectrum (the second summand in the braces) have been separated. Here f_{n0} are the oscillator strengths, $G_{n0}(\omega')$ is the spectral shape of the line for the transition $|0\rangle \to |n\rangle$, and $g_{c0}(\omega')$ is the spectral function of dipole excitation to the continuous energy spectrum. The right-hand side of (2.20) involves $\sigma_{n0}(\omega')$, the cross-section for photoexcitation of the discrete power level, and $\sigma_{c0}(\omega')$, the cross-section for excitation of a state in the continuous spectrum. In view of (2.20), the following expression for the total probability of photoexcitation of the nth state of the discrete spectrum can be obtained from (2.19):

$$W_{n0} = \frac{c}{(2\,\pi)^2}\int_0^\infty \sigma_{n0}(\omega')\,\frac{|\mathbf{E}(\omega')|^2}{\hbar\,\omega'}\,d\omega' \tag{2.21}$$

or

$$W_{n0} = \frac{e^2}{2m} f_{n0} \int_0^\infty G_{n0}(\omega') \frac{|\mathbf{E}(\omega')|^2}{\hbar\,\omega'}\,d\omega'. \tag{2.22}$$

Equations (2.21) and (2.22) are valid within the range of applicability of the perturbation theory, when $W_{n0} < 1$.

When the spectral pulse width $\Delta\omega \propto 1/\Delta t$ is much greater than the spectral width of the bound–bound transition $\Delta\omega \gg \Delta\omega_{n0}$, the line shape function can be replaced by the delta function $G_{n0}(\omega') \to \delta(\omega_{n0} - \omega')$. With this replacement, and in view of the expression for the oscillator strength $f_{n0} = 2m\,\omega_{n0}|\mathbf{d}_{n0}|^2 / 3\hbar\,e^2$ (2.21) yields a simple expression for the phototransition probability:

$$W_{n0} = |\Omega(\omega_{n0})|^2, \tag{2.23}$$

where $\Omega(\omega_{n0}) = \mathbf{d}_{n0}\,\mathbf{E}(\omega_{n0})/\hbar$ is the Fourier transform of the instantaneous Rabi frequency calculated at the eigenfrequency of the excited transition. Naturally, (2.23) can be obtained directly from the electric field strength to first order of the perturbation theory using (2.4) and (2.5).

For the probability of excitation of an arbitrary state in the continuous spectrum over the whole time of action of a field pulse, (2.19) and (2.20) imply

$$W_{c0} = \frac{c}{(2\pi)^2} \int_{I_P/\hbar}^\infty \sigma_{c0}(\omega') \frac{|\mathbf{E}(\omega')|^2}{\hbar\,\omega'}\,d\omega', \tag{2.24}$$

where I_P is the ionization potential of the system under consideration. For the differential probability of photoexcitation to a state of the continuous spectrum with energy ε, (2.24) yields

$$\frac{dW}{d\omega_{\varepsilon 0}} = \frac{c}{(2\pi)^2} \sigma_{c0}(\omega_{\varepsilon 0}) \frac{|\mathbf{E}(\omega_{\varepsilon 0})|^2}{\hbar\,\omega_{\varepsilon 0}}, \tag{2.25}$$

where $\omega_{\varepsilon 0} = (I_P + \varepsilon)/\hbar$ is the frequency of transition to a specified state of the continuous spectrum. The formula (2.25) is obtained from (2.24) by substituting the delta function $\delta(\omega_{\varepsilon 0} - \omega')$ into the integrand on the right-hand side of (2.24). This substitution separates out the transition of the system from the ground state to a state of the continuous spectrum with a specified energy $|0\rangle \to |\varepsilon\rangle$.

For an *arbitrary* photoinduced process in a quasi-monochromatic field, one can obtain the following expression for the probability per unit time in terms of the process cross-section:

$$w_{tot}(t) = \int_0^\infty \sigma(\omega') \frac{I(\omega', t)}{\hbar\,\omega'} d\omega'. \tag{2.26}$$

Here the cross-section $\sigma(\omega')$ can describe not only photoabsorption, but also other photoprocesses, such as scattering, stimulated bremsstrahlung, and stimulated photorecombination. Integrating (2.26) with respect to time, the probability for the whole time of action of the pulse is

$$W = \int_{-\infty}^{\infty} w(t)\, dt = \int_0^\infty \sigma(\omega') \int_{-\infty}^{\infty} \frac{I(\omega', t)}{\hbar\,\omega'} dt\, d\omega'. \tag{2.27}$$

Hence in view of (2.11), we arrive at the formula (2.19).

For a quasi–monochromatic field, analogously to (2.21)–(2.25), the expression for the electromagnetic pulse energy absorbed by the quantum system over the whole time of interaction with the radiation is

$$\Delta \mathrm{E} = \int_{-\infty}^{\infty} Q(t)\, dt = \int_0^\infty \sigma(\omega') \int_{-\infty}^{\infty} I(\omega', t) dt\, d\omega'. \tag{2.28}$$

Hence in view of (2.11), we find

$$\Delta \mathrm{E} = \frac{c}{(2\,\pi)^2} \int_0^\infty \sigma(\omega') \left|\mathbf{E}(\omega')\right|^2 d\omega'. \tag{2.29}$$

Thus (2.19) and (2.25) can be used to describe not only photoabsorption in an ultrashort pulse, but also other photoprocesses, such as radiation scattering and the stimulated bremsstrahlung effect. For this purpose, a suitable cross-section must be substituted in the right-hand side of (2.19).

Let us consider photoexcitation of the system under the action of a laser pulse with a Gaussian envelope i.e., with electric field strength

$$\mathbf{E}(t) = \mathbf{E}_0 \exp\left(-t^2/\Delta t^2\right) \cos(\omega\, t + \varphi), \tag{2.30}$$

where $\mathbf{E}_0$ is the amplitude, ω is the carrier frequency, φ is the CE phase, and Δt is the parameter proportional to the pulse duration Δt_p.

The Fourier transform of the field (2.30) has the form

$$E(\omega') = E_0 \frac{\sqrt{\pi}}{2} \Delta t \left\{ \exp\left[-i\,\phi - \frac{(\omega - \omega')^2 \Delta t^2}{4}\right] + \exp\left[i\,\phi - \frac{(\omega + \omega')^2 \Delta t^2}{4}\right] \right\}. \tag{2.31}$$

It is natural to define the pulse duration Δt_p as the ratio of the probability W for the whole time of action of the pulse to the probability per unit time w for the same pulse shape in the limit of long duration $\left(\Delta t_p \gg \omega^{-1}\right)$:

$$\Delta t_p \leftarrow \frac{W}{w}. \tag{2.32}$$

Then for the field (2.30), the relationship between the parameter Δt and the pulse duration Δt_p is

$$\Delta t_p = \sqrt{\frac{\pi}{2}}\Delta t \cong 1.253\,\Delta t. \tag{2.33}$$

To derive (2.33), we took the limit $\Delta t \to \infty$ in the Fourier transform of the pulse (2.31): $|\mathbf{E}(\omega')|^2 \to (\pi/2)^{3/2}\,\mathbf{E}_0^2\,\Delta t\,\delta(\omega - \omega')$, whence (2.12) leads to $W_{tot} \to \sqrt{\pi/2}\,(K(\omega)/4\hbar^2)\mathbf{E}_0^2\,\Delta t$. Comparing the last relation with the expression $w = \left(K(\omega)/4\hbar^2\right)\mathbf{E}_0^2$ for the photoexcitation rate in a monochromatic field, (2.32) then implies (2.33).

We now consider the number n_c of cycles in a pulse:

$$n_c = \frac{\Delta t_p}{T} = \frac{\omega\,\Delta t_p}{2\,\pi}, \tag{2.34}$$

where $T = 2\,\pi/\omega$ is the period of oscillation at the carrier frequency. Using (2.33), we obtain the following expression for the parameter Δt appearing in the formulas (2.30)–(2.31) in terms of the number of cycles in a pulse:

$$\Delta t = \frac{2\,\sqrt{2\,\pi}\,n_c}{\omega}. \tag{2.35}$$

Hereafter we will consider ultrashort pulses, for which the number of cycles at the carrier frequency is $n_c \geq 1$ and less.

Expressed in terms of the number of cycles, the squared magnitude of the Fourier transform of the field (2.31), which according to (2.19) defines the photoexcitation probability for the whole time of action of the radiation, has the form

$$|E(\omega')|^2 = 2\pi^2\left(\frac{n_c\,E_0}{\omega}\right)^2 G_E(\omega', \omega, n_c)\left[1 + K_{ph}(\omega', \omega, n_c)\,\cos(2\,\varphi)\right], \tag{2.36}$$

where (compare with (1.36) and (1.37) in Chap. 1)

$$G_E(\omega', \omega, n_c) = \exp\left[-4\,\pi\,n_c^2\left(1 - \frac{\omega'}{\omega}\right)^2\right] + \exp\left[-4\,\pi\,n_c^2\left(1 + \frac{\omega'}{\omega}\right)^2\right], \tag{2.37}$$

$$K_{ph}(\omega', \omega, n_c) = \operatorname{sech}\left(8\,\pi\,n_c^2\,\frac{\omega'}{\omega}\right). \tag{2.38}$$

For the probability of excitation of the discrete power level for the whole time of action of the pulse, (2.22) and (2.36) yield

$$W_{n0} = 2\pi^2\left(\frac{d_{n0}\,E_0}{\hbar\omega}\,n_c\right)^2 G_E(\omega_{n0}, \omega, n_c)\left[1 + K_{ph}(\omega_{n0}, \omega, n_c)\,\cos(2\,\varphi)\right] \tag{2.39}$$

under the assumption that $W_{n0} < 1$. We see here that the function $G_E(\omega_{n0}, \omega, n_c)$ describes the spectral form of an excitation line of the bound–bound transition, while $K_{ph}(\omega_{n0}, \omega, n_c)$ is the phase modulation factor since it specifies the dependence of the process probability on the CE phase.

Note that, for a pulse of the form (1.26a), instead of the equality (2.37), one should use (1.36a), taking into account the relation (2.35) between the parameters Δt and n_c.

From the resulting expression (2.39) it follows that, under the conditions of validity of the perturbation theory, the dependence of the total probability of photoexcitation of the bound–bound transition under the action of a pulse of the form (2.30) on the CE phase is given by the function $\cos(2\,\varphi)$. From numerical analysis of the right-hand side of (2.38), it follows that the phase modulation factor has an appreciable value only for subcycle pulses: $n_c < 0.5$. For a fixed value of the parameter n_c, the factor K_{ph} grows with carrier frequency ω, and in this case, according to the expression for the spectral function of excitation (2.37), the process probability decreases.

In the limit of a pulse (2.30) of *zero duration* ($\Delta t \to 0$), when $E(\omega') \to \sqrt{\pi}\,E_0\,\Delta t\,\cos\varphi$, (2.19) takes the form

$$W_{tot} = \frac{c}{4\,\pi\,\hbar}\,(E_0\,\Delta t)^2\cos^2\varphi\int_0^\infty \frac{\sigma(\omega')}{\omega'}\,d\omega'. \tag{2.40}$$

Thus in this case the total photoabsorption probability is defined by the frequency integral of the ratio $\sigma(\omega')/\omega'$. In this case the carrier envelope (CE) phase dependence is given by the function $\cos^2\varphi$, which corresponds to the formula (2.39), since $K_{ph} \to 1$ as $\Delta t \to 0$.

Note that, for excitation by a pulse of the form (1.26a), Eq. (2.40) should be rewritten as

$$W_{tot} = \frac{c}{16\,\pi\,\hbar}\,E_0^2\,\Delta t^6\,\sin^2\varphi\int_0^\infty \sigma(\omega')\,\omega'^3\,d\omega'. \tag{2.40a}$$

For the total absorbed energy over the whole time of action of the pulse and in the limit $\Delta t \to 0$, (2.29) implies

$$\Delta\mathrm{E} = \frac{\pi}{2} N_e r_e (c\,\Delta t)^2 E_0^2 \cos^2\varphi, \tag{2.41}$$

where $r_e = e^2/m c^2$ is the classical electron radius and N_e is the number of electrons in the atom. To derive (2.41), we used the sum rule for the photoabsorption cross-section:

$$\int_0^\infty \sigma(\omega')\,d\omega' = \frac{2\pi^2 e^2}{m c} N_e. \tag{2.42}$$

Equations (2.40)–(2.41) are valid within the framework of applicability of the perturbation theory, when $W_{tot} < 1$. It follows from (2.19) that the limit of zero pulse duration is realized while satisfying $\Delta t < 1/\Delta\omega_a$, where $\Delta\omega_a$ is the frequency interval giving the main contribution to the process cross-section integrated with respect to the frequency.

Thus the formulas (2.19) and (2.27) derived in this section express the total photoprocess probability for the whole time of action of the radiation in terms of the cross-section of this process and the Fourier transform of the electric field strength in the pulse. The resulting equations describe a photoprocess induced by an ultrashort electromagnetic pulse, when the concepts of probability per unit time and radiation intensity are inapplicable but it is nevertheless possible to use the perturbation theory.

2.2 Excitation of a Substance Under the Action of Ultrashort Pulses

2.2.1 Phase Control of Photoexcitation by an Ultrashort Laser Pulse

We now use (2.19) obtained in the previous section to calculate the photoexcitation of a multielectron atom by an ultrashort Gaussian pulse of radiation (2.30) in the local plasma frequency model. Within the framework of this model the expression for the photoabsorption cross-section of the atom is [3]

$$\sigma_{ph}^{(BL)}(\omega') = \frac{2\pi^2 e^2}{m c} \int n(r)\,\delta\big(\omega' - \omega_{pl}(r)\big)\,d\mathbf{r}, \tag{2.43}$$

where $\omega_{pl}(r) = \sqrt{4\pi e^2 n(r)/m}$ is the local plasma frequency and $n(r)$ is the spatial distribution of electron density in the atom. Substituting (2.43) into (2.19), we find

$$W_{tot}^{(ph)} = \frac{\sqrt{\pi}\, e}{\sqrt{m}\, \hbar} \int_0^{\infty} \left| E\left(\omega_{pl}(r), \varphi\right)\right|^2 \sqrt{n(r)}\, r^2\, dr, \tag{2.44}$$

where $\left|E\left(\omega_{pl}(r), \varphi\right)\right|^2$ is the squared magnitude of the Fourier transform of the electric field in a pulse of the form (2.31), calculated at the local plasma frequency, in which the CE phase dependence is explicitly specified. To analyze the phase effects in the total probability of photoexcitation by ultrashort laser pulses, we introduce the phase modulation factor by the formula

$$K_{tot}^{(ph)} = 2\,\frac{W_{tot}^{(ph)}(\varphi = 0) - W_{tot}^{(ph)}(\varphi = \pi/2)}{W_{tot}^{(ph)}(\varphi = 0) + W_{tot}^{(ph)}(\varphi = \pi/2)}. \tag{2.45}$$

The phase modulation factor of the total probability of photoabsorption by an atom with charge $Z = 30$ calculated using the statistical model for the electron density is presented in Fig. 2.1 for three pulse widths as a function of the carrier frequency. It will be recalled that the dimensionless parameter n_c is the number of periods at the carrier frequency in the radiation pulse. We note that appreciable dependence of the photoabsorption probability on the CE phase occurs only for $n_c < 0.5$, and also in the case of a bound–bound transition. The phase modulation factor for the fixed parameter n_c grows with carrier frequency. It should be noted that the probability of photoabsorption at the high-frequency boundary of the interval presented in Fig. 2.1 is 15 % of its maximum value, corresponding in this model to the frequency $\omega_{\max} = 0.4$ a.u.

The expression (2.19) for the total photoabsorption probability can be applied to the interaction of an ultrashort pulse with a *metal nanosphere* in a dielectric medium. When the radiation wavelength is much longer than the nanoparticle radius r_s, its dynamic polarizability can be described by the Lorentz formula

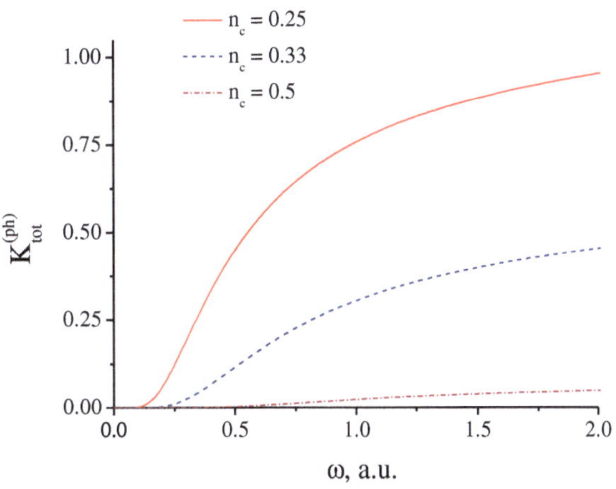

Fig. 2.1 Phase modulation factor for the total probability of photoabsorption of an ultrashort pulse by an atom as a function of the carrier frequency

$$\beta_s(\omega) = \varepsilon_m \frac{\varepsilon_s(\omega) - \varepsilon_m}{\varepsilon_s(\omega) + 2\,\varepsilon_m} r_s^3, \tag{2.46}$$

where $\varepsilon_s\,(\omega)$ is the dielectric permittivity of the nanoparticle metal and ε_m is the dielectric permittivity of the matrix. Hence we can use the optical theorem to find the photoabsorption cross-section in the dipole approximation and then (2.19) to calculate the total photoabsorption probability for the whole time of action of the pulse.

The results of calculations to find the probability of photoabsorption of an ultrashort pulse by a silver nanoparticle in a glass matrix are shown in Fig. 2.2 for two values of the CE phase. The frequency dependence of the dielectric permittivity of silver is restored using data for the real and imaginary parts of the refractive index.

We thus that in this case ($n_c = 0.25$) photoabsorption depends heavily on the CE phase, especially for photon energies at carrier frequencies exceeding the maximum energy. With increasing pulse duration, the phase dependence of the probability becomes less noticeable, and for $n_c > 0.5$ it practically disappears.

In a number of cases quantum systems are excited by a sequence of identical pulses separated by some time interval T (not to be confused with the oscillation period). It is not difficult to obtain a Fourier transform of the electric field strength for such a sequence consisting of N identical pulses in terms of the Fourier transform of a single pulse $E(\omega')$:

$$E_N(\omega') = \frac{\sin(\omega' T N/2)}{\sin(\omega' T/2)} \exp\left[i\frac{(N-1)\,\omega' T}{2}\right] E(\omega'). \tag{2.47}$$

Substituting (2.47) into the right-hand side of (2.19), we find the probability of photoexcitation of a quantum transition under the action of N identical pulses:

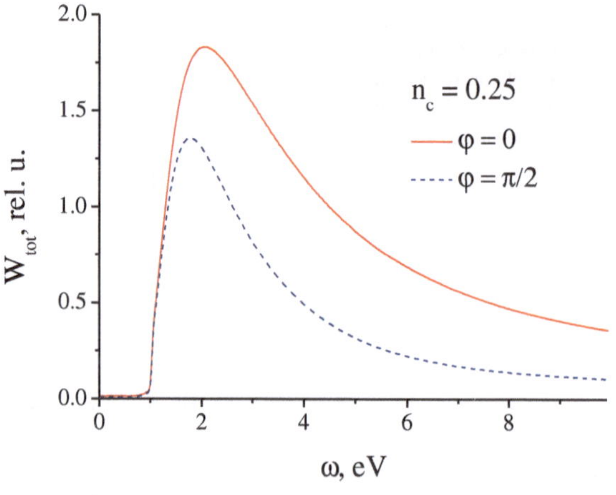

Fig. 2.2 Total probability of photoabsorption of an ultrashort pulse ($n_c = 0.25$) on a silver sphere ($r_s = 5.3$ nm) as a function of the carrier frequency for two values of the CE phase

$$W_{21}(N) = \frac{c}{4\pi^2 \hbar} \int \frac{\sigma_{21}(\omega')}{\omega'} \left[\frac{\sin(\omega' T N/2)}{\sin(\omega' T/2)}\right]^2 |E(\omega')|^2 \, d\omega'. \tag{2.48}$$

We now use these expressions to describe photoionization of a hydrogen atom under the action of a series of short pulses. In this case the process cross-section $\sigma_{21}(\omega)$ is given by the Sommerfeld formula

$$\sigma_H\left(\nu = \frac{\hbar\omega}{Ry}\right) = \frac{2^9 \pi^2 a_B^2}{3 \cdot 137 \cdot \nu^4} \frac{\exp\left(-\frac{4\, arctg\sqrt{\nu-1}}{\sqrt{\nu-1}}\right)}{1 - \exp\left(-2\pi/\sqrt{\nu - 1}\right)}, \tag{2.49}$$

where $a_B \cong 0.53$ Å is the Bohr radius and $R_y \cong 13.6$ eV denotes the Rydberg energy. Figure 2.3 shows the results of calculations using (2.48) and (2.49) to find the probability of photoionization of a hydrogen atom by a series of laser pulses with a width of two oscillation periods at the carrier frequency. Plotted on the abscissa is the value $\nu = \varepsilon/Ry + 1$, where $\varepsilon = \hbar\omega - Ry$ is the energy of an ionized electron. Clearly, with a growing number of pulses, the spectral dependence of the photoexcitation probability narrows near the maximum value determined from the equation $\omega T = 2\pi k$ (where k is a natural number). Since the energy of an electron knocked out of an atom is $Ry\,(\nu - 1)$ and the number of these electrons is proportional to the probability $W(N)$, the figure shows that the energy spectrum of the photoelectrons can be controlled by changing the parameters of the series of exciting pulses.

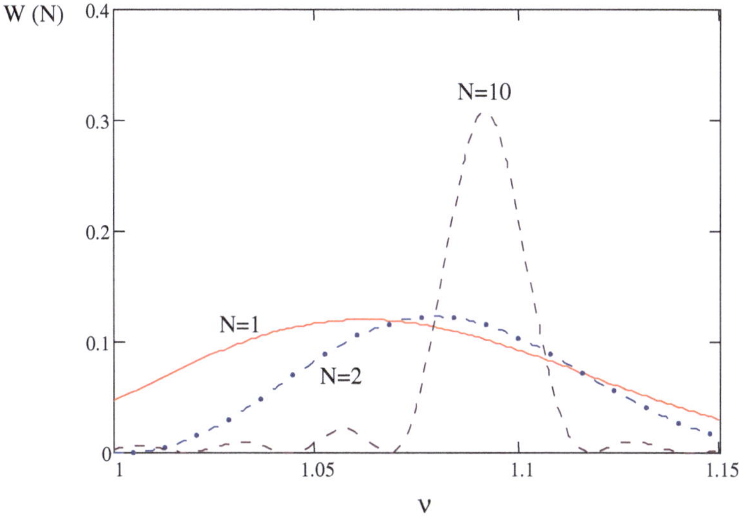

Fig. 2.3 Probability of photoionization of a hydrogen atom under the action of a sequence of N two-cycle laser pulses, $\nu = \hbar\omega/Ry$

2.3 Dependence of the Excitation Probability on the Duration of Ultrashort Pulses

We now consider the dependence of the photoexcitation probability on the duration of ultrashort Gaussian pulses (2.30) for the example of a multielectron atom in the Lenz–Jensen statistical model [4]. For the photoabsorption cross-section we use the local plasma frequency model (2.43). Then the process probability is determined by the basic formula (2.19) and the Fourier transform of the electric field strength in the pulse.

2.3.1 Atom in the Lenz–Jensen Statistical Model

Common to all statistical models of an atom, the expression for the local electron density has the form

$$n_{LJ}(r) = Z^2 f(x = r/r_{TF})\, a_B^{-3}, \tag{2.50}$$

$$r_{TF} = \frac{b}{\sqrt[3]{Z}} a_B, \quad b = \sqrt[3]{\frac{9\,\pi^2}{128}} \cong 0.8853$$

where Z is the charge on the atomic nucleus, r_{TF} is the Thomas–Fermi radius, $a_B = \frac{\hbar^2}{me^2}$ is the Bohr radius, and $f(x)$ is an universal function of the dimensionless distance to the nucleus $x = r/r_{TF}$ depending on the specific statistical model of the atom. Lenz and Jensen proposed the following expression for this function:

$$f_{LJ}(x) \cong 3.7e^{-\sqrt{9.7x}} \frac{\left(1 + 0.26\sqrt{9.7x}\right)^3}{(9.7x)^{3/2}}. \tag{2.51}$$

This provides a more realistic description of the electron density distribution in a multielectron atom than the Thomas–Fermi function. The spectral photoabsorption cross-section of an atom in the Lenz–Jensen model calculated using (2.43), (2.50), and (2.51) is presented in Fig. 2.4 for two values of the charge on the atomic nucleus.

2.3.2 Total Photoabsorption Cross-Section

We now write down the expression for the total photoabsorption cross-section (2.19) separating out the dependence on the duration of the ultrashort pulses as expressed through the number of cycles n_c [see the determination of the number of cycles (2.34)]:

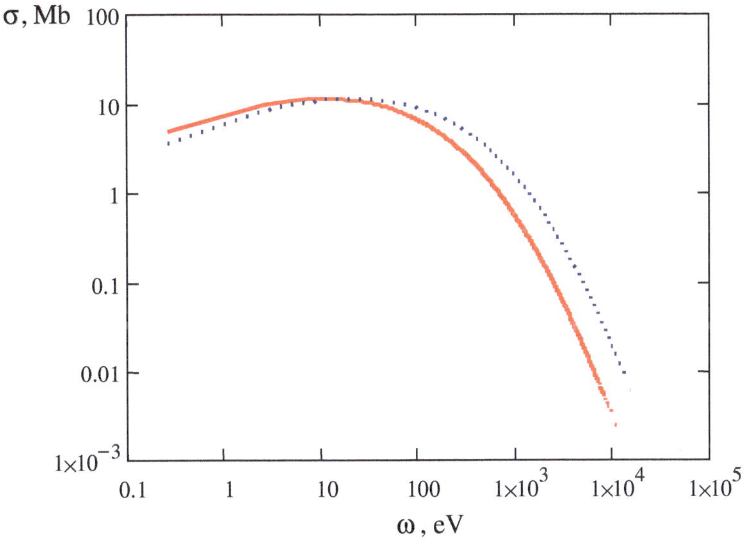

Fig. 2.4 Spectral photoabsorption cross-section of a Lenz–Jensen atom calculated in the local plasma frequency model for two nuclear charges: *solid line* Z = 30, *dotted line* Z = 60

$$W_{tot}(n_c) = \frac{c}{(2\pi)^2} \int_0^\infty \sigma(\omega) \frac{|\mathbf{E}(\omega, n_c)|^2}{\hbar\omega} d\omega, \tag{2.52}$$

where $\sigma(\omega)$ is the spectral cross-section of the process, c is the velocity of light, and $\mathbf{E}(\omega, n_c)$ is the Fourier transform of the electric field strength in the pulse.

The advantage of ultrashort pulse width measurements in terms of the number of pulses in the expression for the total photoabsorption probability consists in the universal nature of the resulting dependences, which hold over a wide spectral range.

Once normalized (to the squared magnitude of the Fourier transform of the electric field strength peak value $|\mathbf{E}_0(\omega)|^2$), the total photoprocess probability in the monochromatic limit $n_c \gg 1$ can be written

$$\tilde{W}_{tot}^{(mon)} = \frac{W_{tot}^{(mon)}}{|\mathbf{E}_0(\omega)|^2} \to \frac{c\,\sigma(\omega)}{\hbar\,\omega^2} n_c. \tag{2.53}$$

Thus the photoprocess probability in the monochromatic limit (the long pulse limit) grows linearly with the number of cycles n_c in the pulse, as expected from traditional considerations. Moreover, in the monochromatic limit the photoprocess probability does not depend on the CE phase φ.

In the opposite limiting case of a pulse of zero width ($n_c \ll 1$) and constant CE phase ($\Phi = \varphi = const$), it can be shown that (2.36)–(2.38) lead to

$$W_{tot}(n_c) = \frac{4\pi}{3}\left(\frac{E_0}{\hbar\omega}\right)^2 n_c^2 \langle 0|\hat{\mathbf{d}}^2|0\rangle \cos^2\varphi, \tag{2.54}$$

where $\langle 0|\hat{\mathbf{d}}^2|0\rangle$ is the quantum–mechanical mean value of the squared magnitude of the atomic electron electric dipole moment operator and E_0 is the electric field amplitude in the pulse. The sum rule was used to derive the right-hand side of (2.54).

Thus in the limit of a pulse of zero width the total photoprocess probability is proportional to the *squared* number of cycles in the pulse and the squared cosine of the CE phase.

Here we are interested in the dependence of the total photoprocess probability on the parameter n_c in the low value region $n_c \sim 1$ and $n_c \leq 1$. To determine this dependence, we use the above formulas describing photoabsorption in the framework of the local plasma frequency model, and with the atomic electron density in the Lenz–Jensen model. The results of our calculations are given in Figs. 2.5, 2.6, 2.7, and 2.8 for a Lenz–Jensen atom with nuclear charge $Z = 30$ and normalized total photoabsorption probability $\tilde{W}_{tot}$ determined by

$$\tilde{W}_{tot} = \frac{W_{tot}}{|\mathbf{E}_0|^2}. \tag{2.55}$$

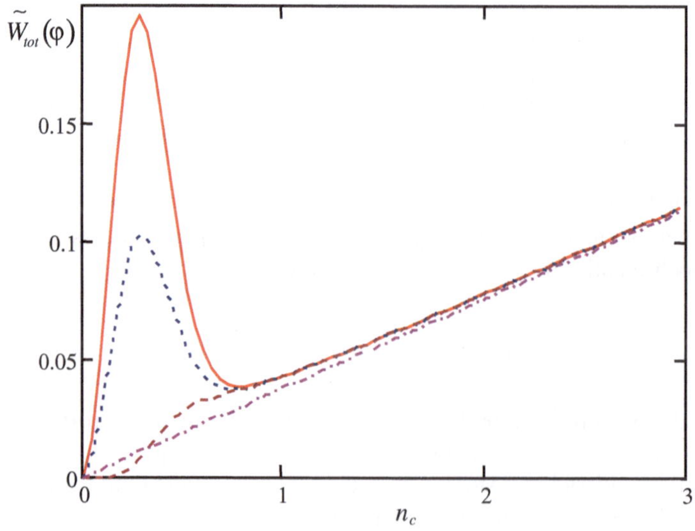

Fig. 2.5 Total photoabsorption probability for ultrashort pulses with different values of the CE phase, calculated for a Lenz–Jensen atom: *solid line* $\varphi = 0$, *dotted line* $\varphi = \pi/4$, *dashed line* $\varphi = \pi/2$, *dash-and-dot line*—monochromatic limit, $\hbar\omega = 10$ a.u., Z = 30

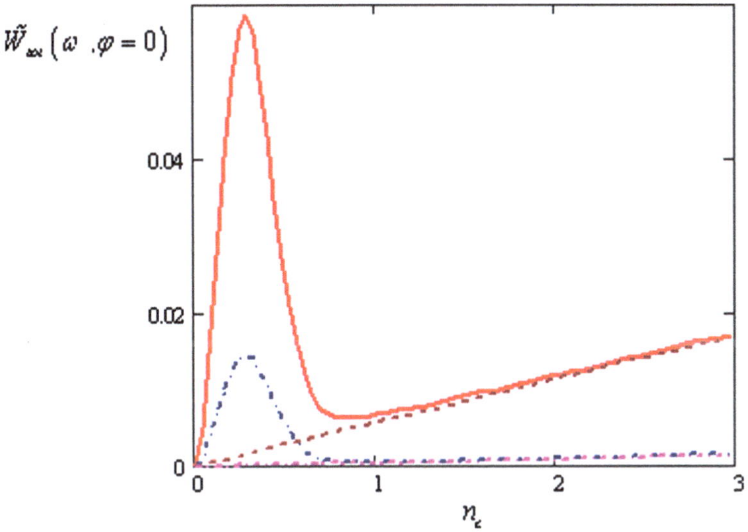

Fig. 2.6 Total photoabsorption probability for ultrashort pulses with the CE phase equal to zero, at two different carrier frequencies and for a Lenz–Jensen atom: *solid line* $\hbar\omega = 500$ eV, *dash-and-dot line* $\hbar\omega = 1000$ eV. *Straight lines* show corresponding monochromatic limits

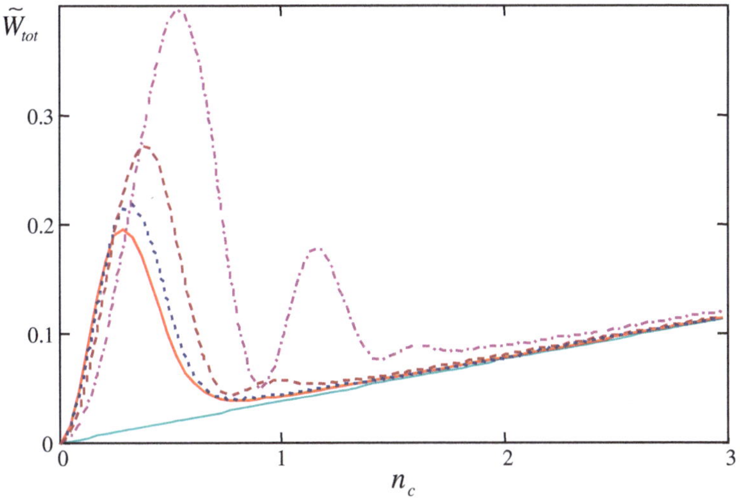

Fig. 2.7 Total photoabsorption probability for chirped ultrashort pulses and for a Lenz–Jensen atom, $\hbar\omega = 10$ a.u., $Z = 30$: *solid red line* $\alpha = 0$, *dotted line* $\alpha = 0.5$, *dashed line* $\alpha = 1$, *dash-and-dot line* $\alpha = 2$, *solid cyan line*—monochromatic limit

The probability of photoabsorption under the action of ultrashort pulses with constant CE phase $\Phi(t) = \varphi = \text{const}$ is shown in Fig. 2.5 for a specified value of the carrier frequency and different values of the parameter φ.

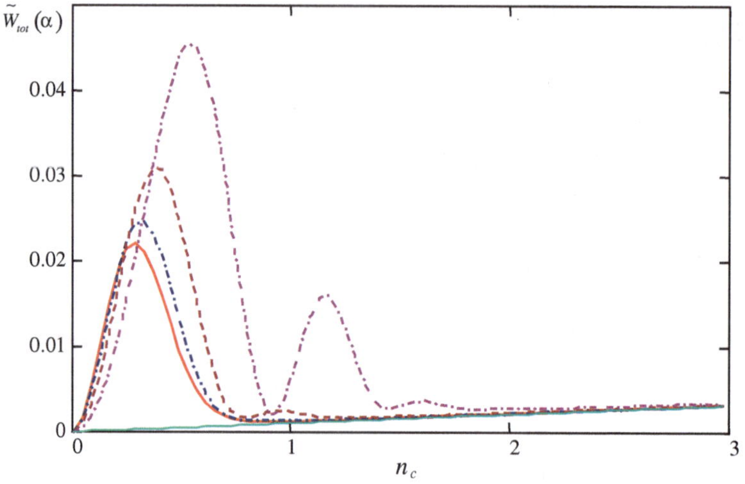

Fig. 2.8 Total photoabsorption probability for chirped ultrashort pulses and for a Lenz–Jensen atom, $\hbar\omega = 30$ a.u., Z = 30: *solid red line* $\alpha = 0$, *dotted line* $\alpha = 0.5$, *dashed line* $\alpha = 1$, *dash-and-dot line* $\alpha = 2$, *solid cyan line*—monochromatic limit

For subcycle pulses, it can be seen that the probability depends strongly on the CE phase. For example, for a cosine pulse ($\varphi = 0$), photoabsorption has a maximum at $n_c \leq 0.5$. The amplitude of this maximum decreases with growing CE phase.

The straight line in Figs. 2.5, 2.6, 2.7, and 2.8 represents the photoabsorption probability in the monochromatic limit, when the function $\tilde{W}(n_c)$ has linear behavior [see (2.53)]. As can be seen from Fig. 2.5, the total photoabsorption probability tends to a straight line for $n_c \geq 1$, as it should in the monochromatic limit. Thus in the case considered, the probability $\tilde{W}(n_c)$ depends nonlinearly on the pulse width (the parameter n_c) only for subcycle ultrashort pulses.

Figure 2.6 shows the dependence of the total probability of atomic photoabsorption on the parameter n_c for a cosine pulse ($\varphi = 0$) and different carrier frequencies ω expressed in electron-volts. From this figure it follows in particular that, for the case under consideration, the function $\tilde{W}(n_c)$ has a smaller angle of inclination to the X axis for higher values of the carrier frequency.

2.3.3 Excitation of an Atom by a Chirped Pulse

Let us consider the dependence of the photoexcitation probability on the duration of ultrashort pulses when an atom is acted on by a chirped electromagnetic pulse and the initial phase Φ is quadratically time-dependent: $\Phi(t) = kt^2$. The Fourier transform of the electric field strength in such a pulse is given by (1.34).

The results of the corresponding calculations are shown in Figs. 2.7 and 2.8 for different values of the carrier frequency in the Lenz–Jensen model.

We see that the function $\tilde{W}(n_c)$ is in this case somewhat more complex than for a fixed CE phase (compare with Figs. 2.5 and 2.6). For example, for high enough values of the dimensionless chirp α there are several maxima of the photoabsorption probability as a function of the number of cycles in a pulse, and the value of these maxima decreases with growing parameter n_c. It is significant that for a chirped pulse the photoabsorption probability maxima occur not only for subcycle ultrashort pulses, but also in the region $n_c > 1$.

In the case of a multicycle chirped pulse, coincidence of the process probability with the monochromatic limit is retained for high enough values of the parameter n_c. However, the number of cycles in a pulse at which this coincidence begins to occur depends on the value of the dimensionless chirp $\alpha = \kappa \Delta t^2$ (see (1.34)). The higher the chirp value, the more cycles are required in the pulse to reach the monochromatic limit (linear dependence of the probability $\tilde{W}(n_c)$ on the parameter n_c).

From comparison of Figs. 2.7 and 2.8 it follows that, as for the constant CE phase, in the monochromatic limit the photoabsorption dependence on the number of cycles in a pulse becomes weaker with growing carrier frequency. This is explained by reduction of the photoabsorption cross-section of a Lenz–Jensen atom with growing frequency in the spectral range $\omega > 10$ eV (see Fig. 2.5).

2.4 Scattering of Ultrashort Pulses by Atoms and in a Plasma

Most works treating the interaction of ultrashort pulses with a material substance are devoted to photoionization and photoexcitation of atomic particles [1]. The description of photoprocesses within the framework of the perturbation theory then has a rather limited domain of applicability since, for characteristic values of the radiation intensity used in modern experiments, nonlinear effects are rather significant.

For radiation scattering by atomic particles, due to the low value of the process cross-section, the perturbation theory turns out to be applicable over a much wider region of parametric variation than in photoionization and photoexcitation. However, application of the usual formulas of perturbation theory obtained in the long pulse limit in the case of single-cycle and subcycle pulses becomes, generally speaking, incorrect. In this section we develop a method to describe scattering of ultrashort electromagnetic pulses by an atom and in a plasma, taking into account possible excitation of the target and the non-dipole nature of the electromagnetic interaction.

2.4.1 Scattering of an Ultrashort Pulse by an Atom

Let us consider scattering of an ultrashort electromagnetic pulse by an atom, taking into account possible excitation of the target [5]. We assume that the spatio-temporal dependence of the electric field strength in the pulse has the form

$$\mathbf{E}(t,\mathbf{r}) = \mathbf{e}\,E_0\,g\left(t - \frac{\mathbf{nr}}{c}\right), \tag{2.56}$$

where $\mathbf{e}$ is the unit polarization vector, E_0 is the field amplitude, $\mathbf{n}$ is the unit vector in the direction of the electromagnetic pulse propagation, $g(\tau)$ is the dimensionless function defined by a concrete realization of the pulse, and c is the velocity of light.

We will decompose the strength (2.56) into plane waves with frequencies ω and wave vectors $\mathbf{k} = (\omega/c)\mathbf{n}$. Then scattering of an electromagnetic field pulse can be represented as scattering of a set of plane waves to a plane wave with frequency ω', unit polarization vector $\mathbf{e}'$, and wave vector $\mathbf{k}' = (\omega'/c)\mathbf{n}'$.

With the above picture, the differential probability of scattering for the whole time of action of the pulse with simultaneous excitation of the target from the state $|i\rangle$ to the state $|f\rangle$ can be obtained in the form

$$\frac{dW_{fi}}{d\Omega'\,d\omega'} = \int_0^\infty \frac{d\sigma_{fi}(\mathbf{k}',\mathbf{k})}{d\Omega'\,d\omega'}\,\frac{dN_{ph}}{d\omega\,dS}\,d\omega, \tag{2.57}$$

where

$$\frac{d\sigma_{fi}(\mathbf{k}',\mathbf{k})}{d\Omega'\,d\omega'} = \delta\left(\omega - \omega' - \omega_{fi}\right)\frac{\omega'^3\,\omega}{c^4}\left|e_l'^{*}\,e_s\,c_{fi}^{ls}(\mathbf{k}',\mathbf{k})\right|^2 \tag{2.58}$$

is the differential scattering cross-section with respect to the solid angle and frequency for a plane wave, with $c_{fi}^{ls}(\mathbf{k}',\mathbf{k})$ the radiation scattering tensor taking into account excitation of the target, and

$$\frac{dN_{ph}}{d\omega\,dS} = \frac{c}{(2\pi)^2}\,\frac{|\mathbf{E}(\omega)|^2}{\hbar\,\omega} \tag{2.59}$$

is the number of photons forming the electromagnetic pulse field in the spectral range $(\omega,\ \omega + d\omega)$ that passed through the unit area during the whole time of action of the radiation, with $\mathbf{E}(\omega)$ the Fourier transform of the electric field strength. Substituting (2.58) into the right-hand side of (2.57), we arrive at an expression for the differential photoprocess probability over the whole time of action of the field that generalizes the expression obtained in [6] to take into account excitation of the target and the non-dipole nature of the electromagnetic interaction.

Gathering the results (2.56)–(2.58), we find the following basic equation:

$$\frac{dW_{fi}}{d\Omega' \, d\omega'} = \frac{\omega'^3}{c^3} \frac{E_0^2}{4\pi^2 \hbar} \left| g\left(\omega' + \omega_{fi}\right) \right|^2 \left| e_l'^{*} \, e_s \, c_{fi}^{ls}(\mathbf{k}', \mathbf{k}) \right|^2, \; \mathbf{k} = \frac{\omega' + \omega_{fi}}{c} \mathbf{n}, \tag{2.60}$$

where $g(\omega)$ is the Fourier transform of the temporal pulse shape function $g(\tau)$ and twice-repeated indices are summed over.

Equation (2.60) can also be obtained using second order perturbation theory in a consistent quantum–mechanical approach [6].

Hereafter we assume that the scattering tensor reduces to the scalar $c_{fi}^{ls} = \delta_{ls} \, c_{fi} \; (c_{fi} = (1/3) \, c_{fi}^{ll})$, with summation over twice-repeated indices. Then on the right-hand side of (2.60) we have a scalar product of polarization vectors of incident and scattered waves, averaged in the usual way for non-polarized scattered radiation.

We consider two cases: (1) scattering without excitation of the target (so-called elastic scattering) and (2) scattering with excitation of the target into an arbitrary state.

For scattering without change of the atomic state and in the multiplicative approximation, we have

$$c_{ii}(\mathbf{k}', \mathbf{k}) \simeq \beta_i(\omega') \, \tilde{F}_{ii}(\mathbf{k}' - \mathbf{k}), \tag{2.61}$$

where $\beta_i(\omega')$ is the dynamic dipole polarizability of an atom in the initial state and $\tilde{F}_{ii}(\mathbf{q}) = F_{ii}(\mathbf{q})/Z$ is the atomic form factor normalized by the number of electrons. In view of (2.61), after summation over polarizations of the scattered photon, (2.60) implies

$$\frac{dW_{ii}}{d\Omega' \, d\omega'} = \frac{1 - (\mathbf{e}\,\mathbf{n}')^2}{4\pi^2} \left(\frac{\omega'}{c}\right)^3 \frac{E_0^2}{\hbar} \left|g(\omega')\right|^2 \left|\beta_i(\omega')\right|^2 \tilde{F}_{ii}^2\left(2\,\frac{\omega'}{c}\,\sin\left(\frac{\theta}{2}\right)\right), \tag{2.62}$$

where θ is the scattering angle. Here, when we write Δk in the argument of the atomic form factor, we take into account the fact that $\omega' = \omega$.

The formula (2.62) can be rewritten in terms of the atomic polarization charge

$$Z_{pol}(\omega) = \frac{m\,\omega^2}{e^2} \left|\beta(\omega)\right| \tag{2.63}$$

as

$$\frac{dW_{ii}}{d\Omega' \, d\omega'} = \frac{1 - (\mathbf{e}\,\mathbf{n}')^2}{4\pi^2} \left(\frac{e^2}{\hbar c}\right)^3 \left(\frac{I_0}{I_a}\right) Z_{pol}^2(\omega') \, \frac{\left|\omega_a \, g(\omega')\right|^2}{\omega'} \, \tilde{F}_{ii}^2\left(2\,\frac{\omega'}{c}\,\sin\left(\frac{\theta}{2}\right)\right), \tag{2.64}$$

where $I_a = c\,\frac{m^4 e^{10}}{8\pi\hbar^8} \simeq 3.5 \cdot 10^{16}\ \mathrm{W/cm^2}$, $\omega_a = \frac{m e^4}{\hbar^3}$ are the atomic units of radiation intensity and frequency, and $I_0 = c\,E_0^2/8\pi$ is the mean radiation intensity. In

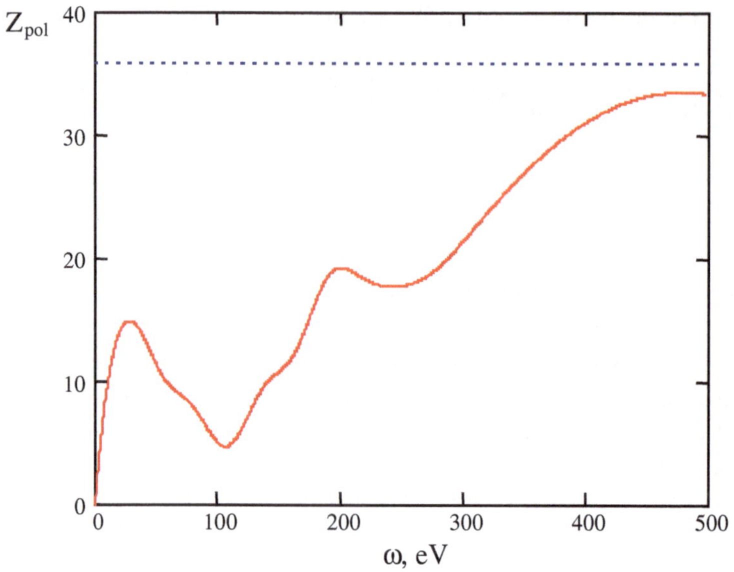

Fig. 2.9 Polarization charge of a krypton atom as a function of frequency. The *dashed line* shows the number of electrons in the atom

the high-frequency limit $\omega \gg \omega_t$ (ω_t is the characteristic eigenfrequency of the target), the polarization charge is equal to the number of electrons in the target.

The frequency dependence of the polarization charge of a krypton atom calculated using experimental photoabsorption data is presented in Fig. 2.9. The solid curve in the figure is obtained by calculating the imaginary part of the polarizability of the krypton atom with the optical theorem and restoring the real part by means of the Kramers–Kronig relation.

When the carrier frequency ω_c of the pulse is close to one of the eigenfrequencies of excitation of the atom in the discrete spectrum $\omega_c \approx \omega_{ri}$ (in this case the oscillator strength for the corresponding transition is nonzero $f_{ri} \neq 0$), the resonant approximation can be used for the polarizability:

$$|\beta_i(\omega' \approx \omega_{ri})|^2 \cong \frac{\pi}{2}\left(\frac{e^2}{m\,\omega'^2}\right)^2 \frac{\omega'}{\delta_{ri}} f_{ri}^2\, \omega'\, G_{ri}(\omega'), \tag{2.65}$$

where δ_{ri} and $G_{ri}(\omega)$ are the width and shape of the resonance transition line. Then instead of (2.62) we have

$$\frac{dW_{ii}(\omega_c \approx \omega_{ri})}{d\Omega'\, d\omega'} = \frac{1-(\mathbf{e}\,\mathbf{n}')^2}{8\,\pi}\left(\frac{e^2}{\hbar\, c}\right)^3 \left(\frac{I_0}{I_a}\right) \frac{\omega'}{\delta_{ri}} |\omega_a\, g(\omega')|^2 f_{ri}^2\, G_{ri}(\omega'). \tag{2.66}$$

In writing (2.66), we have taken into account the fact that, in the spectral range under consideration, the atomic form factor can be assumed equal to unity.

In particular, the resulting formula implies resonant amplification of scattering due to the presence of the multiplier ω'/δ_{ri}.

For an ultrashort pulse with carrier frequency in the optical range, the spectrum is generally much broader than the line of a resonance transition in an atom, so the spectral dependence of the scattering probability will be defined mainly by the shape $G_{ri}(\omega')$ of the spectral line of the transition. In the general case, the spectrum of a scattered pulse will also be influenced by the function $|g(\omega')|^2$.

The resonance probability (2.66) can be generalized to the case where the atom is excited in the scattering process if the frequency of scattered radiation is close to one of the eigenfrequencies for transition of the atom from the intermediate state to the final state.

The probability of scattering with target excitation to an arbitrary state (the entire scattering spectrum) is obtained by summing the probability (2.60) over all possible states $|f\rangle$. Let us consider the entire scattering spectrum in the high-frequency range ($\omega \gg \omega_t$), when the approximate expression for the electromagnetic field scattering tensor is valid:

$$c_{fi}^{(hf)}(\mathbf{k}', \mathbf{k}) \simeq -\frac{e^2}{m\,\omega'\,\omega} F_{fi}(\mathbf{k}' - \mathbf{k}). \tag{2.67}$$

Substituting the right-hand side of (2.67) into the formula (2.60) and summing over all possible final states, we find the following expression for the scattering probability with excitation of the atom:

$$\frac{dW_{tot}^{(hf)}}{d\omega' d\Omega'} = \frac{c\,E_0^2}{4\pi^2\hbar}\left[1 - (\mathbf{e}\,\mathbf{n}')^2\right] r_e^2 \int_0^\infty |g(\omega)|^2 S_i(\omega' - \omega, \mathbf{k}' - \mathbf{k})\,\frac{d\omega}{\omega}, \tag{2.68}$$

where

$$S_i(\Delta\omega, \Delta\mathbf{k}) = \frac{1}{2\pi}\int_{-\infty}^{\infty} dt\, e^{-i\Delta\omega t}\,\langle i|\hat{n}(\Delta\mathbf{k}, t)\hat{n}(-\Delta\mathbf{k})|i\rangle \tag{2.69}$$

is the dynamic form factor (DFF) of an atom in the ith state (see Appendix III).

In the elementary approximation, the DFF of a hydrogen-like atom is

$$S(\Delta\omega, \Delta\mathbf{k}) \simeq \delta\left(\Delta\omega + \frac{\hbar}{2m}\Delta\mathbf{k}^2\right). \tag{2.70}$$

Substituting the right-hand side of (2.70) into (2.68) and integrating with respect to the frequency ω, we find

$$\frac{dW_{tot}^{(hf)}}{d\omega' d\Omega'} \approx \frac{2}{\pi}\left[1 - (\mathbf{e}\,\mathbf{n}')^2\right] \frac{I_0}{\hbar}\, r_e^2\, \frac{|g[\omega(\omega', \theta)]|^2}{\omega(\omega', \theta)}, \tag{2.71}$$

where

$$\omega(\omega',\theta) = \omega_r + \omega' \cos\theta - \omega_r \sqrt{1 - 4\frac{\omega'}{\omega_r}\sin^2\left(\frac{\theta}{2}\right) - \left(\frac{\omega'}{\omega_r}\right)^2 \sin^2(\theta)},$$
$$\omega_r = \frac{m c^2}{\hbar} \tag{2.72}$$

For $\omega' \ll \omega_r$, we have $\omega(\omega',\theta) \cong \omega'$, and instead of (2.71) we obtain

$$\frac{dW_{tot}^{(hf)}}{d\omega' d\Omega'} \approx \frac{1 - (\mathbf{e}\,\mathbf{n}')^2}{4\pi^2}\left(\frac{e^2}{\hbar c}\right)^3 \left(\frac{I_0}{I_a}\right) \frac{|\omega_a\, g(\omega')|^2}{\omega'}. \tag{2.73}$$

This expression describes the entire spectrum for scattering of an ultrashort pulse by a one-electron atom in the high-frequency approximation.

In the high-frequency limit, when $Z_{pol} \cong Z$, the ratio of the probability of "elastic" scattering of an ultrashort pulse by a hydrogen-like atom (2.64) to the total probability (2.73) is equal to the squared normalized form factor of the atom:

$$R^{(hf)}(\omega',\theta) \equiv \frac{dW_{ii}^{(hf)}}{dW_{tot}^{(hf)}} = \tilde{F}_{ii}^2\left(2\frac{\omega'}{c}\sin\left(\frac{\theta}{2}\right)\right). \tag{2.74}$$

Let us consider scattering of an ultrashort Gaussian electromagnetic pulse by an atom. Then the function determining the time dependence of the electric field strength in (2.56) looks like

$$g(\tau) = \exp(-\tau^2/\Delta t^2)\cos(\omega_c \tau + \varphi), \tag{2.75}$$

where ω_c is the carrier frequency, Δt is the time parameter proportional to the pulse width, and φ is the carrier phase with respect to the envelope (the CE phase). It is convenient to express the parameter Δt in terms of the number of periods in the pulse at the carrier frequency n_c: $\Delta t = 2\sqrt{2}\pi n_c/\omega_c$. In view of this fact, the squared magnitude of the Fourier transform of the function (2.75) appearing in the expressions for the scattering probability can be represented as

$$|g(\omega)|^2 = 2\pi^2\left(\frac{n_c}{\omega_c}\right)^2 G_E(\omega,\omega_c,n_c)\left[1 + K_{ph}(\omega,\omega_c,n_c)\cos(2\varphi)\right], \tag{2.76}$$

where

$$G_E(\omega,\omega_c,n_c) = \exp\left[-4\pi n_c^2\left(1 - \frac{\omega}{\omega_c}\right)^2\right] + \exp\left[-4\pi n_c^2\left(1 + \frac{\omega}{\omega_c}\right)^2\right] \tag{2.77}$$

is the spectral form of the pulse and

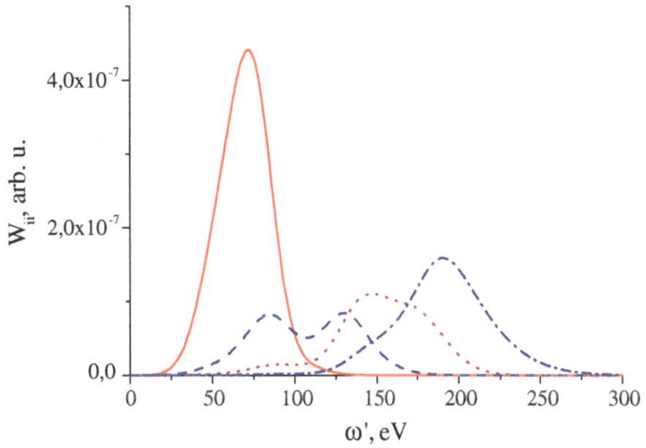

Fig. 2.10 Spectrum of "elastic" scattering of a single-cycle pulse by a krypton atom for different values of the carrier frequency: *solid curve* $\hbar\omega_c = 80$ eV, *dashed curve* $\hbar\omega_c = 110$ eV, *dotted curve* $\hbar\omega_c = 140$ eV, *dash-and-dot curve* $\hbar\omega_c = 180$ eV

$$K_{ph}(\omega, \omega_c, n_c) = \operatorname{sech}\left(8\,\pi\,n_c^2\,\frac{\omega}{\omega_c}\right) \tag{2.78}$$

is effectively the phase modulation factor. It follows from (2.78) that the phase modulation factor has an appreciable value only for ultrashort pulses, when $n_c \sim 1$.

Figure 2.10 presents the spectrum of "elastic" scattering (i.e., without target excitation) of a single-cycle pulse by a krypton atom, as calculated using (2.62) for a scattering angle of 45° and several values of the carrier frequency.

As can be seen from this figure, in the case of a single-cycle pulse, the form of the scattered radiation spectrum essentially depends on the carrier frequency value. Far from the minimum of the frequency dependence of the polarization charge of a krypton atom, which falls approximately on 107 eV (see Fig. 2.9), the spectral scattering curves have a symmetric form with a maximum at the centre. Near the minimum frequency ($\hbar\omega_c = 110$ eV), a dip appears in the scattering spectrum. It follows from (2.64) that the described evolution of the scattering spectrum is explained by superposition of two frequency dependences: one is the ultrashort pulse spectrum (2.76) and the other the atomic polarization charge spectrum. For a single-cycle pulse with spectral width broad enough to be comparable with the scale of spectral singularities of the krypton atom polarization charge, this superposition modifies the form of the spectral scattering curve. The situation changes in going to longer pulses, for example, to a three-cycle pulse, whose spectrum for scattering by a krypton atom is presented in Fig. 2.11 for different carrier frequencies.

We see that for a three-cycle pulse the scattered radiation spectrum is a bell-shaped curve. Its shape is defined by the spectrum of the incident pulse (2.76) and the amplitude depends on the value of the polarization charge of the atom at the carrier frequency.

Figure 2.12 shows the result of calculations using the expression (2.74) for the ratio of the probability of "elastic" scattering of a single-cycle pulse by a hydrogen atom in the high-frequency limit summed over all scattered radiation frequencies to the analogous value for the process with arbitrary excitation of the atom.

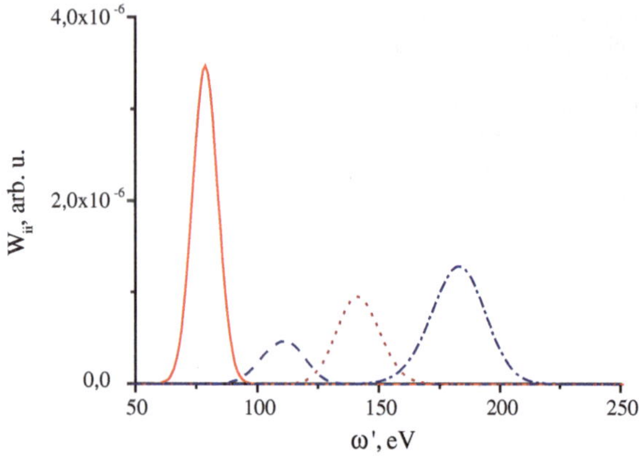

Fig. 2.11 The same as Fig. 2.10 but for a three-cycle pulse

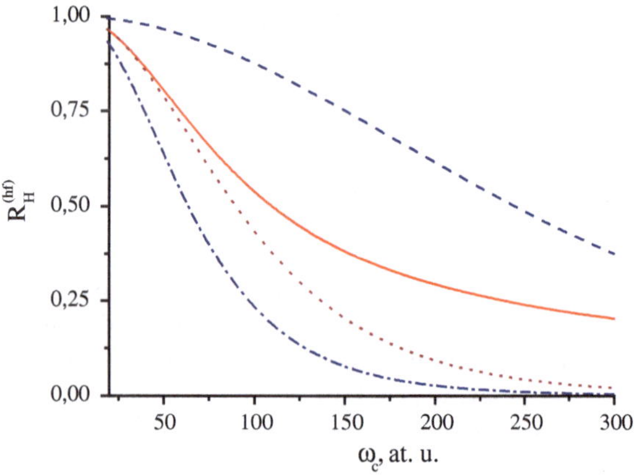

Fig. 2.12 Ratio of the probability of "elastic" scattering of a single-cycle pulse by a hydrogen atom to the total probability calculated in the high-frequency limit for three scattering angles and for probabilities integrated with respect to the angles (*solid curve*): *dashed curve* $\theta = 30°$, *dotted curve* $\theta = 90°$, *dash-and-dot curve* $\theta = 180°$

This ratio is calculated as a function of the carrier frequency of the pulse for different scattering angles. Shown in the same figure by a solid curve is the contribution of the "elastic" process summed over the scattering angles. From Fig. 2.12 we see that, for low values of the carrier frequency of the electromagnetic pulse, scattering occurs mainly without excitation of the atom. For wide scattering angles, the role of the elastic channel decreases more rapidly with growing frequency. For the process probability integrated with respect to the angles, the contributions of the elastic and inelastic channels are compared for a carrier frequency of about 112 a.u., which corresponds to a photon energy of about 3 keV.

2.4.2 *Scattering of an Ultrashort Pulse in a Plasma*

Here we use the approach developed for an atomic target to describe scattering of an ultrashort pulse in a plasma [7]. We proceed from the formula (2.68), obtained in the high-frequency approximation for an atom. It will be recalled that, since plasma electrons are free, the high-frequency condition $\omega \gg \omega_t = 0$ (ω_t is the characteristic eigenfrequency of a plasma electron) is satisfied automatically for them, so (2.68) is applicable. In this expression the dynamic form factor of an atom should be replaced by the dynamic form factor of an electron component of the plasma. This is given by (A.3) and (A.19) in Appendix III.

Scattering of electromagnetic radiation in a plasma can be of two types: Compton scattering and transition scattering. Compton scattering corresponds to large variations of the wave vector $|\Delta\mathbf{k}| > r_D^{-1}$ ($\Delta\mathbf{k} = \mathbf{k}' - \mathbf{k}$), when the electromagnetic interaction proceeds in the single-particle regime. This means that the energy–momentum excess during scattering is transferred to one plasma electron as in Compton scattering of X-radiation by an atom, when an atomic electron is knocked out of the atom, taking away the energy–momentum excess. Corresponding to Compton scattering in the plasma is the first summand on the right-hand side of (A.19), describing the normalized electron dynamic form factor.

For transition scattering, the situation is quite the opposite: the inequation $|\Delta\mathbf{k}| < r_D^{-1}$ is satisfied, implying that scattering of an electromagnetic field occurs by a Debye sphere surrounding an ion in the plasma as by an unit. In this case the energy–momentum excess is transferred to the plasma ion.

Using (2.68) and the explicit expression for the plasma DFF (see Appendix III), then averaging over the polarization of incident radiation and integrating with respect to the scattered radiation frequency, we obtain the following angular distribution (in terms of one ion) for the probability of transition scattering of an ultrashort pulse:

$$\frac{dW_i}{N_i\, d\Omega'} \simeq \frac{c\, E_0^2\, r_e^2}{8\,\pi^2}\, Z_i^2\, (1 + \cos^2\theta) \int_0^\infty \frac{|g(\omega')|^2\, d\omega'}{\hbar\,\omega' \left[1 + (2\, r_D\, (\omega'/c)\, \sin(\theta/2))^2\right]^2}, \tag{2.79}$$

where $N_i = n_i\, \delta V$ is the number of ions in the scattering volume δV.

The analogous expression for the Compton scattering channel in terms of one-electron looks like

$$\frac{dW_e}{N_e\, d\Omega'} \simeq \frac{c\, E_0^2\, r_e^2}{8\,\pi^2}\, (1 + \cos^2\theta) \int_0^\infty \frac{(2\, r_D\, (\omega'/c)\, \sin(\theta/2))^4 |g(\omega')|^2\, d\omega'}{\hbar\,\omega' \left[1 + (2\, r_D\, (\omega'/c)\, \sin(\theta/2))^2\right]^2}, \tag{2.80}$$

where $N_e = n_e \delta V$. In deriving (2.79)–(2.80), we made the replacement

$$\frac{\exp\left(-\frac{\Delta\omega^2}{2\,\Delta\mathbf{k}^2\, \mathrm{v}_T^2}\right)}{\sqrt{2\pi}\, \mathrm{v}_T\, |\Delta\mathbf{k}|} \to \delta(\Delta\omega), \tag{2.81}$$

which is justified if $|\Delta\omega| \gg; \mathrm{v}_T\, |\Delta\mathbf{k}|$, as supposed here. The relation (2.81) amounts to equating the scattered frequency ω' with the frequency ω in the Fourier expansion of the ultrashort pulse ($\omega' \cong \omega$), that is, it neglects inelastic processes during scattering.

For the total probability of scattering of an ultrashort pulse in a plasma by the transition channel, integrating the right-hand side of (2.79) with respect to the solid angle of scattering, we find

$$\frac{W_i}{N_i} \simeq \frac{c\,E_0^2\,r_e^2}{4\,\pi}\,Z_i^2 \int\limits_0^\infty F_i\left[2\left(r_D\,(\omega'/c)\right)^2\right]\frac{|g(\omega')|^2\,d\omega'}{\hbar\,\omega'}, \tag{2.82}$$

$$F_i(x) = 2\,\frac{1+x}{x^3}\left\{\frac{2x(1+x)}{2x+1} - \ln(2x+1)\right\}.$$

The analogous expression for Compton scattering is

$$\frac{W_e}{N_e} \simeq \frac{c\,E_0^2\,r_e^2}{4\,\pi} \int\limits_0^\infty F_e\left[2\left(r_D\,(\omega'/c)\right)^2\right]\frac{|g(\omega')|^2\,d\omega'}{\hbar\,\omega'}, \tag{2.83}$$

$$F_e(x) = 2\,\frac{2x\,(4x^3+11x^2+15x+6) - (4x^3+8x^2+7x+2)\,\ln(2x+1)}{3x^3\,(2x+1)}.$$

Let us use these expressions to describe scattering of an ultrashort Gaussian pulse (2.75) in a plasma. The value of $|g(\omega')|^2$ to be included in the above expressions is given in this case by (2.76)–(2.78). In the long pulse limit $n_c \gg 1$, (2.76)–(2.78) imply

$$|g(\omega)|^2 \to \pi^2\,\frac{n_c}{\omega_c}\,\delta(\omega - \omega_c). \tag{2.84}$$

In this case (2.82) and (2.83) simplify, and for the total probability of scattering by both channels normalized by the number of electrons $N_e = Z_i\,N_i$, we have

$$\frac{W(n_c \gg; 1)}{N_e} \simeq \frac{\pi\,c\,E_0^2\,n_c}{4\;\hbar\,\omega_c^2}\,r_e^2\;F\left[2\left(d_e\,(\omega_c/c)\right)^2\right], \tag{2.85}$$

where $F = F_e + Z_i F_i$. This analysis shows that the difference between the probability of scattering of an ultrashort pulse in a plasma as calculated using (2.82) and (2.83) and in the monochromatic limit (2.85) exists only for subcycle pulses $n_c \leq 1$. This difference increases with growing carrier frequency of the pulse and has a non-monotonic dependence on the Debye radius.

The results of calculations of the spectral and angular probability of scattering of an electromagnetic pulse in plasma are illustrated in Figs. 2.13, 2.14, 2.15, and 2.16. Plotted on the ordinate in these figures is the scattering probability for the whole time of action of a pulse normalized to the intensity $I_0 = c\,E_0^2/8\,\pi$. We assume everywhere that $\omega_c > \omega_{pe}$, where ω_{pe} is the electron plasma frequency. The dependence of the spectral curve of scattering of ultrashort pulses on the CE

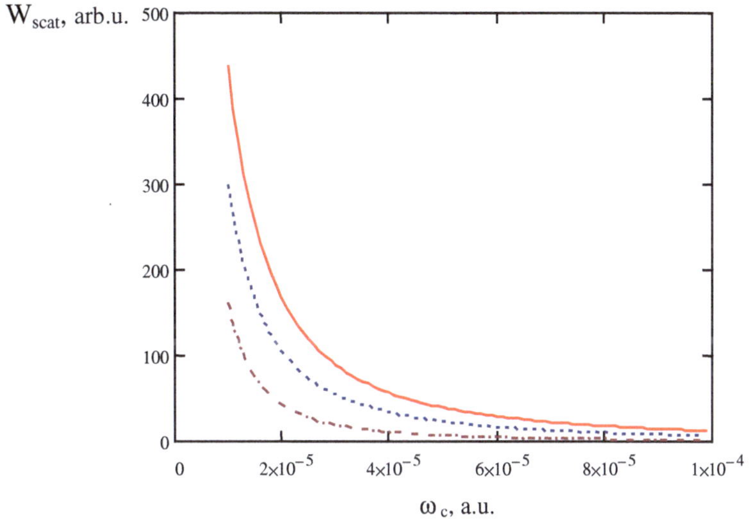

Fig. 2.13 Dependence of the normalized probability of scattering of a quarter-cycle pulse in a plasma on the carrier frequency for three values of the CE phase, $r_D = 10^5$ a.u. *solid curve* $\varphi = 0$, *dotted curve* $\varphi = \pi/4$, *dash-and-dot curve* $\varphi = \pi/2$

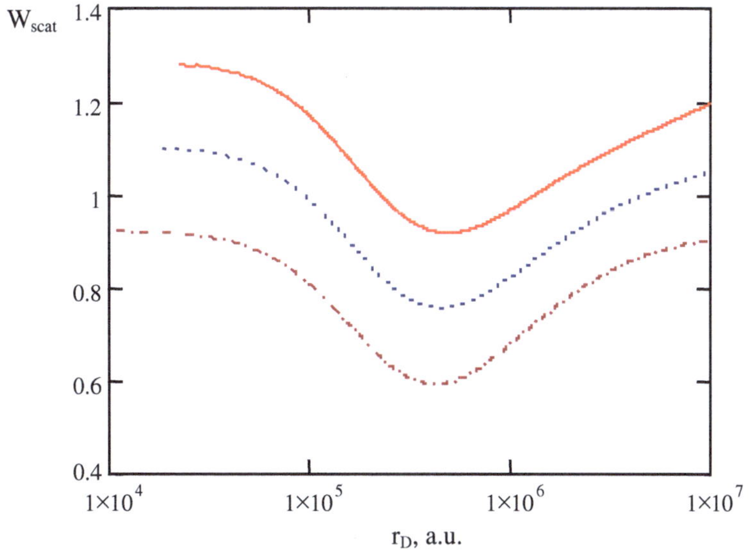

Fig. 2.14 Normalized probability of scattering of a half-cycle pulse as a function of the Debye radius for different values of the CE phase, $\omega_c = 3{\cdot}10^{-4}$ a.u. *solid curve* $\varphi = 0$, *dotted curve* $\varphi = \pi/4$, *dash-and-dot curve* $\varphi = \pi/2$

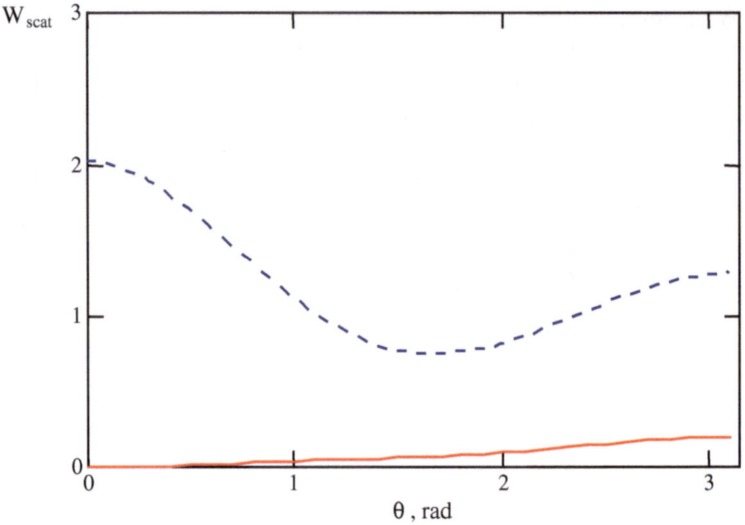

Fig. 2.15 Angular dependence of the normalized probability of scattering of a quarter-cycle pulse ($\omega_c = 10^{-4}$ a.u., $r_D = 10^6$ a.u.) by different channels: *solid curve* compton scattering, *dotted curve* transition scattering for $Z_i = 1$

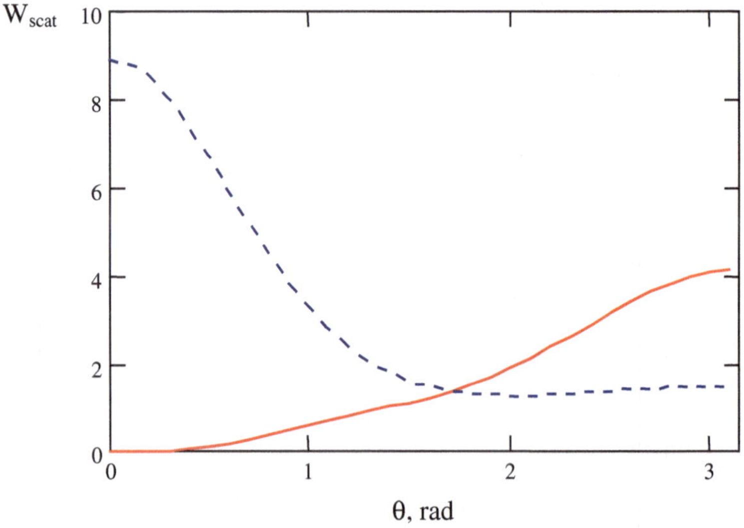

Fig. 2.16 The same as in Fig. 2.15 but for a five-cycle pulse

phase is manifest only for subcycle pulses $n_c < 1$. In this case, as can be seen from Fig. 2.13, growth of the CE phase in an interval from 0 to $\pi/2$ results in decreasing scattering probability.

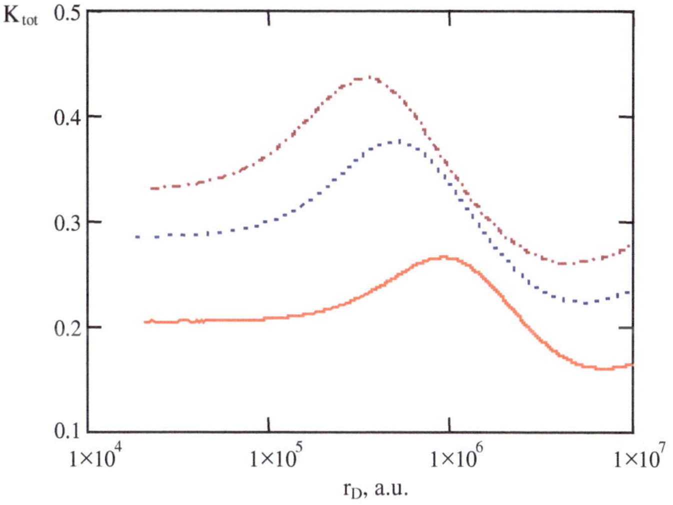

Fig. 2.17 Dependence of the phase modulation factor on the Debye radius for different values of the carrier frequency of a half-cycle pulse: *solid curve* $\omega_c = 10^{-4}$ a.u., *dotted curve* $\omega_c = 2{\cdot}10^{-4}$ a.u., *dash-and-dot curve* $\omega_c = 3{\cdot}10^{-4}$ a.u

The dependence of the scattering probability for the whole time of action of a half–cycle pulse on the value of the Debye radius is presented in Fig. 2.14 for the carrier frequency $\omega_c = 3{\cdot}10^{-4}$ a.u. and three values of the CE phase. It can be seen that corresponding curves have a minimum near the value $r_D \approx 5{\cdot}10^5$ a.u. The calculation shows that, with decreasing carrier frequency, the minimum in the dependence of the scattering probability on the Debye radius is shifted to the region of greater values. With growing pulse duration, this minimum is weakly shifted to the region of smaller Debye radii.

Figures 2.15 and 2.16 show the values for the two process channels as a function of the angle of photon scattering for quarter-cycle (Fig. 2.15) and five-cycle (Fig. 2.16) pulses.

We see that with increasing angle the probability of Compton scattering grows, and the probability of transition scattering has a non-monotonic dependence.

From Figs. 2.15 and 2.16, it follows that, with decreasing pulse width, the relative contribution of Compton scattering of an ultrashort electromagnetic pulse in a plasma decreases in comparison with the contribution of transition scattering.

To characterize the dependence of the probability of ultrashort pulse scattering on the CE phase, it is convenient to introduce the phase modulation factor for the scattering probability by the formula

$$K_{tot} = 2\frac{W(\varphi = 0) - W(\varphi = \pi/2)}{W(\varphi = 0) + W(\varphi = \pi/2)}. \quad (2.86)$$

The results for calculations of this value for scattering of a half-cycle pulse in a plasma are shown in Figs. 2.17 and 2.18. Figure 2.17 graphs the dependence of the

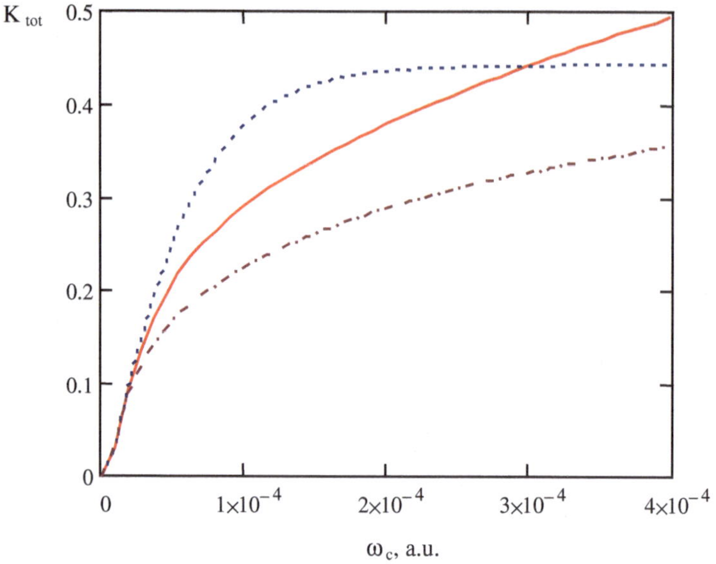

Fig. 2.18 Dependence of the phase modulation factor for scattering of a half-cycle pulse on the carrier frequency for different values of the Debye radius: *solid curve* $r_D = 10^5$ a.u., *dotted curve* $r_D = 10^6$ a.u., *dash-and-dot curve* $r_D = 10^7$ a.u

phase modulation factor (2.86) on the value of the Debye radius for different values of the carrier frequency of a pulse. It is found that this dependence has non-monotonic behavior, the maximum being shifted to the region of smaller radii with increasing carrier frequency. Furthermore, from Fig. 2.17 we see that the phase dependence of the scattering probability grows with increasing carrier frequency of the ultrashort pulse. The same fact follows from the plots of Fig. 2.18, in which the phase modulation factor is presented as a function of the carrier frequency for different values of the Debye radius.

The above analysis thus shows that the dependence of the probability of scattering of an ultrashort electromagnetic pulse on the Debye radius in plasma has non-monotonic behavior. The minimum of the pulse scattering probability is shifted to the region of smaller values of the Debye radius with decreasing carrier frequency and increasing pulse duration.

From the analysis of angular dependences of two channels of scattering of an ultrashort pulse in plasma, it follows that, with decreasing pulse duration, the contribution of Compton scattering to the total probability of the process decreases, and in this case the probability of transition scattering for wide angles increases.

It is found that the phase dependence of scattering is manifest only for subcycle pulses $n_c < 1$, with growth of the CE phase in an interval from 0 to $\pi/2$ resulting in decreasing process probability, and increased carrier frequency resulting in increasing phase modulation factor (2.86).

A.1 2.5 Appendix III Dynamic Form Factor of Plasma Particles

The dynamic form factor (DFF) defines the probability of electromagnetic interactions with participation of plasma particles, during which the subsystem of plasma electrons or ions absorbs the energy–momentum excess. Such processes are exemplified by radiation scattering in plasma, bremsstrahlung and polarization bremsstrahlung on plasma particles including the stimulated bremsstrahlung effect, and a number of other phenomena.

The determination of the DFF of a specified plasma component has the form

$$S(\omega, \mathbf{k}) = \frac{1}{2\pi} \int_{-\infty}^{\infty} dt\, e^{i\,\omega t} \langle \hat{n}(\mathbf{k}, t)\, \hat{n}(-\mathbf{k}) \rangle, \tag{A.1}$$

where $\hat{n}(\mathbf{k})$, $\hat{n}(\mathbf{k}, t)$ are spatial Fourier transforms of the operators representing the concentration of plasma particles of a specified type in the Schrödinger and Heisenberg pictures, and the angle brackets include both quantum–mechanical and statistical averages.

It will be recalled that the Heisenberg representation of quantum–mechanical operators takes into account their time dependence, in contrast to the Schrödinger representation, in which the whole time dependence is transferred to the wave function of the system. The relationship between these representations for an arbitrary operator $\hat{Q}$ is

$$\hat{Q}(t) = \exp\left(i \hat{\mathrm{H}} t/\hbar\right) \hat{Q} \exp\left(-i \hat{\mathrm{H}} t/\hbar\right),$$

where $\hat{\mathrm{H}}$ is the Hamiltonian of the quantum–mechanical system. In this appendix, however, the quantum–mechanical formalism will not be used, and the quantum description is given only for completeness. Equation (A.1) can be obtained from the formula

$$S(\omega, \mathbf{k}) = \sum_{f,i} w(i)\, \delta\left(\omega + \omega_{fi}\right) \left|n_{fi}(\mathbf{k})\right|^2, \tag{A.2}$$

averaging over initial states $|i\rangle$ and summing over final states $|f\rangle$ of the plasma particles ($w(i)$ is the probability of a plasma particle being in the ith state). As usual the delta function in (A.2) reflects energy conservation.

Depending on the type of plasma particles, the DFF can be electronic, ionic, or mixed. For the mixed DFF, the product of the density operators for electrons and ions appears in the determination of (A.1).

Physically, the DFF defines the probability of plasma absorption of the four-dimensional wave vector $k = (\omega, \mathbf{k})$ in terms of the action of an external disturbance *on a specified plasma component*. When the charge distribution in the plasma is uniform, this probability is equal to zero, since then the Fourier transform of the density of charged particles reduces to the delta function $n(\mathbf{k}) \to n\,\delta(\mathbf{k})$. Thus the DFF is connected with charge fluctuations in the plasma.

In fact, the dynamic form factor reflects the dynamics of plasma particles interacting with each other through long-range Coulomb forces. Interactions are then taken into account within the ensemble of one type of particles and also between electrons and ions.

For an uniform plasma, it is convenient to introduce the DFF of the unit volume (the normalized DFF) by the formula

$$\tilde{S}(\omega, \mathbf{k}) = \frac{S(\omega, \mathbf{k})}{V}, \tag{A.3}$$

where V is the volume of the plasma. This equation follows from the fact that, for an uniform medium, the pair correlation function of the concentration depends only on the relative distance between spatial points:

$$Kn(\mathbf{r}, \mathbf{r}', t) \equiv \langle \hat{n}(\mathbf{r}, t)\,\hat{n}(\mathbf{r}', 0)\rangle = Kn(\mathbf{r} - \mathbf{r}', t).$$

To calculate the normalized DFF, it is convenient to use the fluctuation–dissipation theorem connecting the DFF of a plasma component with the function describing the plasma response to the external electromagnetic disturbance [8]. This theorem for the electron DFF is expressed by

$$\tilde{S}_e(\omega, \mathbf{k}) = \frac{\hbar}{\pi e^2} \frac{\mathrm{Im}\{F_{ee}(\omega, \mathbf{k})\}}{[\exp(-\hbar\omega/T) - 1]} \tag{A.4}$$

where $F_{ee}(\omega, \mathbf{k})$ is the linear function describing the electron component response to the fictitious external potential acting only on plasma electrons, and T is the temperature of the plasma in energy units. The imaginary part of the response function appearing in (A.4) describes energy dissipation in the plasma, whence the name for the theorem.

With reference to [8], we introduce a second linear function describing the response to the external potential $F_{ei}(\omega, \mathbf{k})$, i.e., describing the response of the electron component of the plasma to the action of the fictitious external potential acting only on plasma ions. Here for convenience we use the Coulomb gauge of the electromagnetic field, in which the divergence of the vector potential is equal to zero ($\mathrm{div}\mathbf{A} = 0$) and the charge density is related only to the scalar potential of the electromagnetic field φ via the Poisson equation. So the external potential $\varphi_{ext}(k)$ acts on the plasma, where $k = (\omega, \mathbf{k})$ is the four-dimensional wave vector. Then the density of the electron charge induced in the plasma is expressed in terms of the above response functions as follows:

$$\langle \hat{\rho}_e(k)\rangle = [F_{ee}(k) + F_{ei}(k)]\,\varphi_{ext}(k). \tag{A.5}$$

$\langle \hat{\rho}_j(k)\rangle = e_j \langle \hat{n}_j(k)\rangle$ is the charge density of the jth type of plasma particles. Equation (A.5) indicates that the electron charge density arises in the plasma due to direct action of the external potential on plasma electrons [the first summand in the square brackets of (A.5)] and also as a result of the action of the external potential on plasma ions that are bound to electrons by Coulomb forces. If interaction between particles of type i and type j is weak, one can use the technique described in [8] to express F_{ij} in terms of the characteristics of noninteracting particles. For this purpose the new response function $\alpha_j(k)$ is introduced—the response function for particles of type j to the *total potential* in the plasma. It takes into account the action on charged particles of the potential $\varphi_{ind}(k)$ induced in the plasma due to redistribution of the charged particles under the action of the external potential. With the help of the function $\alpha_j(k)$, the induced charge density for the jth component can be expressed in terms of the total potential:

$$\langle \hat{\rho}_j(k) \rangle = \alpha_j(k)\, \varphi_{tot}(k). \tag{A.6}$$

As the response function $\alpha_j(k)$ describes the action of the total potential on the plasma particles, the characteristics of noninteracting particles can be used to calculate it, since interaction between them is already taken into account in the total potential. This technique is widely used in plasma physics to describe screening and initiation of collective excitations. In the approach under consideration, corresponding to the random phase approximation [8], $\alpha_j(k)$ can be expressed in terms of the function $Q_j(k)$ characterizing the noninteracting particles according to $\alpha_j = e_j^2 Q_j$, where

$$Q_j(k) = \int \frac{n_j(\mathbf{p} + \hbar\mathbf{k}) - n_j(\mathbf{p})}{\mathrm{E}_j(\mathbf{p} + \hbar\mathbf{k}) - \mathrm{E}_j(\mathbf{p}) - \hbar\omega - i0} \frac{2\, d\mathbf{p}}{(2\pi\hbar)^3}. \tag{A.7}$$

Here $n_j(\mathbf{p})$ is the dimensionless momentum distribution function of plasma particles of type j and $\mathrm{E}_j(\mathbf{p}) = p^2/2m_j$. Hereafter we need to know the imaginary part of the function $Q_j(k)$, which can be determined from (A.7) using the Sokhotsky formula. For the Maxwell velocity distribution of the electrons, we find

$$\mathrm{Im}\{Q_j(k)\} = \pi\left(e^{-\hbar\omega/T} - 1\right) n_j \frac{\exp\left\{-\omega^2/2k^2 \mathrm{v}_{Tj}^2\right\}}{\sqrt{2\pi}\, k\, \mathrm{v}_{Tj}}. \tag{A.8}$$

The functions introduced above to describe the response to the total potential are related to the longitudinal part of the dielectric permittivity by

$$\varepsilon^{(l,j)}(k) = 1 - \frac{4\pi}{\mathbf{k}^2} \alpha_j(k). \tag{A.9}$$

We can now solve the original problem, i.e., we will find the function $F_{ee}(\omega, \mathbf{k})$ and express it in terms of the function describing the response to the total potential. For this purpose we introduce the fictitious external potential φ_{ext}^* acting only on electrons. Then according to the definition of $F_{ee}(\omega, \mathbf{k})$, we have

$$\langle \hat{\rho}_e^*(k) \rangle = F_{ee}(k)\, \varphi_{ext}^*(k). \tag{A.10}$$

On the other hand, $\langle \hat{\rho}_e^* \rangle$ can be expressed in terms of α_e:

$$\langle \hat{\rho}_e^*(k) \rangle = \alpha_e(k) \left[\varphi_{ext}^*(k) + \varphi_{ind}^*(k) \right], \tag{A.11}$$

where φ_{ind}^* is the potential induced under the action of φ_{ext}^*, determined in terms of the density of all plasma charges with the help of the Poisson equation:

$$\varphi_{ind}^*(k) = \frac{4\pi}{\mathbf{k}^2} \left[\langle \hat{\rho}_e^*(k) \rangle + \langle \hat{\rho}_i^*(k) \rangle \right]. \tag{A.12}$$

Here

$$\langle \hat{\rho}_i^*(k) \rangle = \alpha_i(k)\, \varphi_{ind}^*(k), \tag{A.13}$$

Since the potential φ_{ext}^* is assumed to act only on electrons. Solving the system of Eqs. (A.8–A.12), we find the following expression for F_{ee}:

$$F_{ee}(k) = \frac{\alpha_e(k) \left[1 - (4\pi/\mathbf{k}^2)\, \alpha_i(k) \right]}{1 - (4\pi/\mathbf{k}^2) \left[\alpha_e(k) + \alpha_i(k) \right]}. \tag{A.14}$$

Substituting (A.13) into (A.4) and using (A.8) and (A.7), we obtain

$$\tilde{S}_e(k) = \left| \frac{\varepsilon^{l(i)}(k)}{\varepsilon^l(k)} \right|^2 |\delta n_e(k)|^2 + z_i^2 \left| \frac{1 - \varepsilon^{l(e)}(k)}{\varepsilon^l(k)} \right|^2 |\delta n_i(k)|^2, \tag{A.15}$$

where

$$|\delta n_{e,i}(k)|^2 = \frac{n_{e,i}}{\sqrt{2\pi}\, v_{Te}\, |\mathbf{k}|} \exp\left(-\frac{\omega^2}{2\, \mathbf{k}^2\, v_{Te,i}^2} \right) \tag{A.16}$$

are the spatio-temporal Fourier transforms of the squared thermal fluctuations of the electron and ionic components of the plasma calculated for the four-dimensional wave vector $k = (\mathbf{k}, \omega)$. z_i is the charge number of the plasma ions and it is implied that the quasi-neutrality condition is satisfied, so that $n_e = z_i\, n_i$.

The expression for the normalized ionic DFF is found in exactly the same way as the electron DFF. For this purpose, one must make the index replacement $e \rightleftarrows i$ and take into account the fact that, in the denominator of (A.4), the ion charge $e_i = z_i\, e$ now appears. Then we obtain:

$$\tilde{S}_i(k) = \left| \frac{\varepsilon^{l(e)}(k)}{\varepsilon^l(k)} \right|^2 |\delta n_i(k)|^2 + z_i^{-2} \left| \frac{1 - \varepsilon^{l(i)}(k)}{\varepsilon^l(k)} \right|^2 |\delta n_e(k)|^2. \tag{A.17}$$

The mixed normalized DFF is given by

$$\tilde{S}_{ei}(k) = z_i^{-1} \left| \frac{1 - \varepsilon^{l(i)}(k)}{\varepsilon^l(k)} \right|^2 |\delta n_e(k)|^2 + z_i \left| \frac{1 - \varepsilon^{l(e)}(k)}{\varepsilon^l(k)} \right|^2 |\delta n_i(k)|^2, \tag{A.18}$$

which follows from the fluctuation–dissipation theorem (A.4) (with the replacement $e^2 \rightarrow e\,e_i$) and the formula for the linear response function F_{ei} describing the initiation of an electron charge induced by the fictitious potential that acts only on ions. This formula has the form

$$F_{ei}(k) = \frac{\left(4\,\pi/\mathbf{k}^2\right)\alpha_i(k)\alpha_e(k)}{1 - \left(4\,\pi/\mathbf{k}^2\right)\left[\alpha_e(k) + \alpha_i(k)\right]}. \tag{A.19}$$

Equation (A.18) is obtained by analogous reasoning to the deduction of (A.13).

Let us explain the physical meaning of the expression (A.14) for the electron DFF. The first summand is connected with the deficiency of electron charge around the electron density fluctuation, caused by electron–electron repulsion. The second summand in this expression describes the electron charge screening the fluctuation of the ionic plasma component. It results from electron–ion attraction. By analogy, in the expression (A.16) for the ionic DFF, the second summand describes the ionic charge screening the electron density fluctuation, while the first summand describes the deficiency of ionic charge around the ionic fluctuation. Finally, in the formula (A.16) for the mixed DFF, the first summand describes the ionic charge screening the electron density fluctuation, and the second summand describes the electron charge screening the ionic density fluctuation.

Let us consider the explicit form of the electron DFF satisfying the inequations $k\,v_{Te} \gg \omega \gg k\,v_{Ti},\ \omega_{pi}$. Then the low frequency approximation is valid for the longitudinal electron dielectric permittivity of the plasma, and the high-frequency approximation is valid for the ionic component. Using the expressions for the longitudinal part of the dielectric permittivity of the plasma and the formula (A.15), we find

$$\tilde{S}_e(k) \simeq \left(\frac{k^2\,r_{De}^2}{1 + k^2\,r_{De}^2}\right)^2 |\delta n_e(k)|^2 + \frac{z_i^2}{\left(1 + k^2\,r_{De}^2\right)^2}\,|\delta n_i(k)|^2. \tag{A.20}$$

From this formula we see that, in the case of long-wavelength fluctuations, when $k^2\,r_{De}^2 \ll 1$ $(k = 2\,\pi/\lambda)$, the first summand describing the deficiency of electron charge around the electron density fluctuation is small. The second summand connected with electron screening of ionic density fluctuations is large. Hence it follows that, in the long-wavelength limit, the transfer of energy–momentum to the plasma proceeds through the electron charge of the Debye sphere around a plasma ion which reacts in a coherent manner to the electric field. That is, interaction is of a collective nature. In the short-wavelength case $k^2\,r_{De}^2 \gg 1$, the situation is opposite: electromagnetic interaction is realized through excitation of individual plasma electrons, under which the Debye sphere "falls apart" due to the strong spatial non-uniformity of the electric field.

References

1. Krausz, F., Ivanov, M.: Rev. Mod. Phys. **81**, 63 (2009)
2. Mandal, P.K., Speck, A.: Phys. Rev. A. **81**, 013401 (2010)
3. Brandt, W., Lundqvist, S.: Phys. Rev. **139**, A612 (1965)
4. Gombas, P.: Die Statistische Theorie des Atoms und ihre Anwendungen. Springer, Wien (1949)
5. Astapenko, V.A.: JETP **112**, 193 (2011)
6. Astapenko, V.A.: Phys. Lett. A. **374**(13–14), 1585 (2010)
7. Astapenko, V.A.: Plasma Phys. Rep. **37**, 972 (2011)
8. Platzmann, P.M., Wolff, P.A.: Waves and Interactions in Solid State Plasmas. Academic Press, New York (1973)

Chapter 3
Two-Level System in the Field of Ultrashort Electro-Magnetic Pulses

3.1 Optical Bloch Vector Formalism

An elementary quantum object, with which an electromagnetic field can interact, is a system consisting of two energy levels connected by a dipole-allowed transition. Such a system is called a two-level system (TLS) and is characterized by two parameters: the eigenfrequency ω_0 and the dipole moment of a transition $d_0 \neq 0$. The TLS describes, in particular, a transition between bound states of an optical impurity center in a solid (a bound–bound transition) if the resonance condition is fulfilled:

$$|\omega - \omega_0| \leq \Delta\omega, \tag{3.1}$$

where ω is the circular frequency of the electromagnetic field and $\Delta\omega$ is the width of the spectral line of a transition at the optical center. It will be recalled that the spectral line describes the field frequency dependence of the binding force between the electromagnetic field and the transition under consideration.

An example of an optical center in a solid that can be described by a TLS is a nitrogen-vacancy (NV) center in diamond. The structure of this formation is presented in Fig. 3.1.

It should be noted that NV centers in diamond have many applications in various problems of photonics and cryptography, and in biomedical investigations [1].

In the case of a NV center, the two-level system is formed by the ground state ^3A (orbital singlet) and the excited state ^3E (orbital doublet). The transition $^3A \leftrightarrow {}^3E$ is a dipole-allowed transition with an energy of 1.945 eV and a high oscillator strength.

We now investigate this scenario in its most general form. We thus consider the interaction of an electromagnetic radiation pulse with a two-level system with lower level energy E_1 and upper level energy E_2 ($E_2 > E_1$). According to the Bohr postulate, the eigenfrequency is therefore $\omega_0 = (E_2 - E_1)/\hbar$. It is convenient to choose the zero energy level halfway between E_1 and E_2, whence $E_1 = -\hbar\,\omega_0/2$ and $E_2 = \hbar\,\omega_0/2$ (see Fig. 3.2).

V. Astapenko, *Interaction of Ultrashort Electromagnetic Pulses with Matter*, SpringerBriefs in Physics, DOI: 10.1007/978-3-642-35969-9_3,

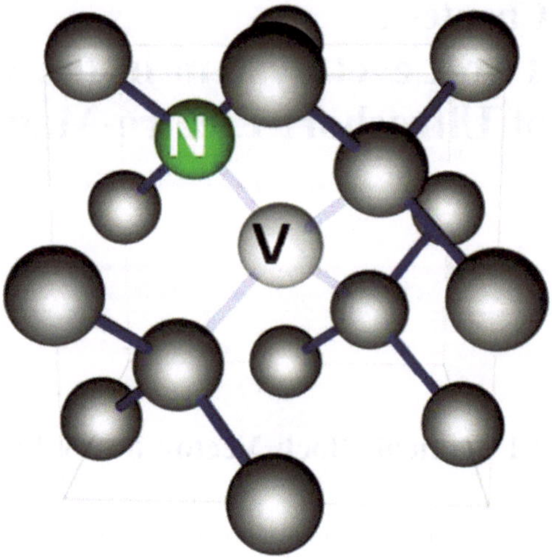

Fig. 3.1 Structure of a NV center in diamond

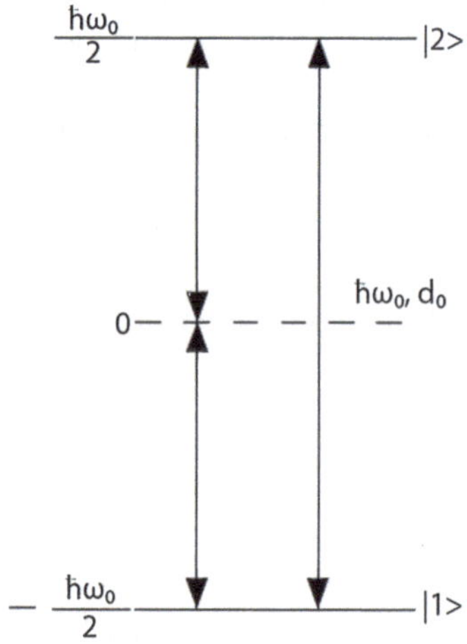

Fig. 3.2 The two-level system

The amplitude of the probability that the system is at the *j*th level is given by the complex number a_j ($j = 1, 2$). The probability amplitudes are related to the level populations N_j by the simple relation $N_j = |a_j|^2$. If interaction of the two-level system with the environment can be neglected, the normalization condition $|a_1|^2+|a_2|^2= 1$ or $N_1 + N_2 = 1$ is satisfied. States of the TLS with a specific energy E_1 or E_2 are called *energy states*, denoted by the Dirac ket vectors $|1\rangle$ and $|2\rangle$. It is known from quantum mechanics that state vectors satisfy the Schrödinger steady-state equation

$$\hat{H}_0 |j\rangle = E_j |j\rangle, \tag{3.2}$$

where $\hat{H}_0$ is the Hamiltonian of the TLS. It will be recalled that the Hamiltonian in the quantum mechanics formalism corresponds to the energy of classical physics. From the mathematical point of view, (3.2) is an eigenvalue equation, with eigenvectors $|j\rangle$ and eigenvalues E_j. It is well known that the eigenvectors of (3.2) must be mutually orthogonal, i.e., $\langle j \mid i \rangle = \delta_{ij}$, where δ_{ij} is the Kronecker symbol. This fact will be used hereafter.

The Dirac vectors (or state vectors) are identical with the wave functions. They completely describe the properties of the quantum system in a given state if its interaction with the environment can be neglected (the pure state). In the general case, at an arbitrary instant of time t, the state of the isolated TLS is specified by a linear superposition of energy states with coefficients a_j:

$$|t\rangle = a_1(t)\,|1\rangle + a_2(t)\,|2\rangle, \tag{3.3}$$

where $|t\rangle$ is the state vector for the TLS at the specified time. It should be emphasized that the whole time dependence is contained in the coefficients a_j, since the energy state eigenvectors $|j\rangle$ are assumed to be time-independent. Thus the time evolution of the TLS is specified by two complex functions $a_j(t)$. If the normalization condition is taken into account, it can be concluded that the state of the TLS at an arbitrary instant of time is described by three real numbers, that is, by some vector in a three-dimensional space. It is convenient to specify this vector in such a way that its components have a clear physical meaning, for example, with the first component describing the dipole moment of the TLS and the third component describing the TLS energy. The state vector $\mathbf{R} = (R_1, R_2, R_3)$ so determined is called the Bloch vector or, in the case of an electric dipole transition, the optical Bloch vector, since it corresponds to electric dipole transitions which have optical eigenfrequencies in atoms.

The components of the Bloch vector are given by the formulas

$$R_1 = a_1\,a_2^* + a_1^*\,a_2 = 2\,\mathrm{Re}\{a_1\,a_2^*\}, \tag{3.4a}$$

$$R_2 = i\left(a_1\,a_2^* - a_1^*\,a_2\right) = -2\,\mathrm{Im}\{a_1\,a_2^*\}, \tag{3.4b}$$

$$R_3 = |a_1|^2 - |a_2|^2. \tag{3.4c}$$

It is easy to check that the third component R_3 describes the mean energy (E) of the two-level system up to a constant factor. According to the quantum mechanics formalism we have

$$\begin{aligned} \mathrm{E} &= \langle t|\hat{H}_0|t\rangle = \left[a_1^*(t)\,\langle 1| + a_2^*(t)\,\langle 2|\right]\hat{H}\left[a_1(t)\,|1\rangle + a_2(t)\,|2\rangle\right] \\ &= \frac{\hbar\,\omega_0}{2}\left[|a_2(t)|^2 - |a_1(t)|^2\right] = -\frac{\hbar\,\omega_0}{2}\,R_3(t). \end{aligned} \tag{3.5}$$

In going from the first line of (3.5) to the second, the orthogonality of the energy state eigenvectors $|j\rangle$ was used.

The dipole moment of the TLS is $d(t) = \langle t|\hat{d}|t\rangle$, where $\hat{d}$ is the dipole moment operator (for simplicity the scalar form of writing is used). We substitute the expression for the state vector (3.3) in this to obtain

$$d(t) \equiv \langle t|\hat{d}|t\rangle = \left(a_1\, a_2^* + a_1^*\, a_2\right) d_0 = R_1\, d_0, \tag{3.6}$$

where $d_0 \equiv d_{21} = \langle 2|\hat{d}|1\rangle$ is the intrinsic dipole moment of the TLS, equal to the matrix element of the dipole moment. In writing (3.6), the equation $\langle 2|\hat{d}|1\rangle = \langle 1|\hat{d}|2\rangle$ was taken into account, along with the fact that, for a spherically symmetric system, $\langle 1|\hat{d}|1\rangle = \langle 2|\hat{d}|2\rangle = 0$.

The second component R_2 is related to the angle φ of rotation of the projection of the Bloch vector on the plane 1×2 (the axes 1, 2, 3 correspond to the projections of the Bloch vector $R_1,\ R_2,\ R_3$) :

$$\varphi = \operatorname{arctg}(R_2/R_1). \tag{3.7}$$

This angle describes the phase of oscillations of the TLS dipole moment. The presence of the second component of the Bloch vector makes it possible to represent the phase of oscillations of the TLS dipole moment as the angle of rotation of the projection of the Bloch vector on the plane 1×2.

Thus the Bloch vector unequivocally determines a state of the TLS, and explicitly specifies its energy and dipole moment (magnitude and phase). From the normalization condition, it follows that the length of the Bloch vector is equal to unity $|\mathbf{R}| = 1$ (with neglected relaxation), whence the end of the Bloch vector circumscribes a sphere of unit radius (the Bloch sphere) (see Fig. 3.3).

It can be shown that, in the case of a free (without external influence) isolated TLS, the Bloch vector rotates around axis 3 with angular velocity equal to the eigenfrequency ω_0 of the two-level system, in the direction from axis 1 to axis 2. This motion is called *free precession* of the Bloch vector. The polar precession

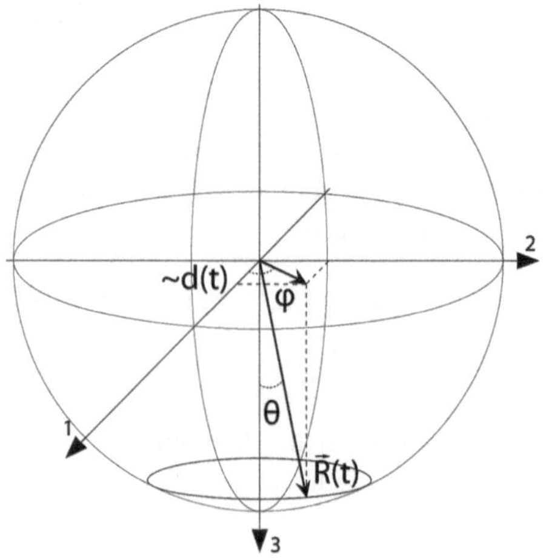

Fig. 3.3 The Bloch vector and the Bloch sphere

angle $\theta = \arccos(R_3)$ is specified by initial conditions as the initial azimuth angle ϕ_0. During free precession, the energy of the two-level system (E) is conserved, while the dipole moment oscillates with frequency ω_0 and magnitude equal to $d_0 \sin\theta$. The polar precession angle is related to the TLS energy through the third projection of the Bloch vector: $\theta = \arccos(-2\,\mathrm{E}/\hbar\,\omega_0)$.

From the above it follows that the dipole moment of the free two-level system is given by

$$d(t) = d_0 \sqrt{1 - R_3^2}\,\cos(\omega_0\,t + \varphi_0). \tag{3.8}$$

Hence we see that the maximum value of the amplitude of the TLS dipole moment is reached for states in which $R_3 = 0$, that is, when the Bloch vector lies in the plane 1×2. Such states are called *coherent states*. In a coherent state the mean energy of the TLS is equal to zero: $\mathrm{E} = 0$ (in view of the chosen zero level). On the other hand, if $R_3 = \pm 1$, the dipole moment of the TLS is equal to zero. The last equation is realized for energy states when $\mathrm{E} = \pm\hbar\,\omega_0/2$ (the Bloch vector is oriented along axis 3). So in energy states, the dipole moment of the two-level system is equal to zero. It will be recalled that the dipole moment defines the intensity of intrinsic emission of the system. Thus in energy states the TLS does not emit, and in coherent states the TLS emission intensity is maximum.

3.1.1 Equation for the Bloch Vector

We now consider interaction of the TLS with an external electric field in the Bloch vector formalism. From the Schrödinger equation for the TLS and the definition of the Bloch vector, it can be shown that

$$\frac{d\mathbf{R}}{dt} = \mathbf{R} \times \mathbf{W}, \tag{3.9}$$

where $\mathbf{W} = (2\,\Omega(t),\, 0,\, \omega_0)$ is the generalized angular velocity vector, $\Omega(t) = d_0\,E(t)/\hbar$ is the time-dependent "instantaneous" Rabi frequency, and $E(t)$ is the electric field strength. From (3.9), it follows that there is a mechanical analogy for an optical transition in an external monochromatic field. The optical Bloch vector is analogous to the angular momentum of a gyroscope: it precesses around the instantaneous direction of the generalized angular velocity $\mathbf{W}(t)$. This angular velocity is defined by the eigenfrequency of the two-level system and the parameters of the external field. In the case of free precession, when $\Omega(t) = 0$, it follows from (3.9) that the Bloch vector rotates counterclockwise around the positive direction of axis 3. The superposition of the external field results in additional rotation around axis 1, with angular velocity $\Omega(t)$. The result is a rather complex motion that is generally difficult to imagine.

3.1.2 Equation for Bloch Vector Motion in the Rotating Wave Approximation

For resonant *monochromatic* radiation or long enough electromagnetic pulses, when the condition (3.1) is satisfied, the pattern of evolution of the Bloch vector in the external field can be considerably simplified if the rotating wave approximation is used. This approximation includes two stages. First, we change to a coordinate system that rotates around axis 3 with angular velocity ω equal to the electromagnetic field frequency in the direction from axis 1 to axis 2. This change is intended to pick up the intrinsic precession of the Bloch vector that occurs when the TLS eigenfrequency ω_0 is close, according to (3.1), to the frequency of the external field. If linearly polarized monochromatic radiation is now represented as the superposition of two circularly polarized waves rotating towards each other, then one of these waves will rotate with the free precession of the Bloch vector (in-phase wave), while the other will rotate in the opposite direction (counterphase wave).

The electric field of the in-phase wave in the rotating coordinate system will be a constant, while the field of the counterphase wave will oscillate at the doubled frequency 2ω. Therefore it is natural to expect the action of the counter-propagating wave on the TLS to be rather limited, and in the rotating wave approximation it is neglected. In view of the above, the equation for the Bloch vector in the rotating coordinate system takes the form

$$\frac{d\mathbf{R}_0}{dt} = \mathbf{R}_0 \times \mathbf{W}_0, \quad \mathbf{W}_0 = (2\Omega_0,\ 0,\ \Delta), \tag{3.10}$$

where $\Omega_0 = d_0 E_0/2\hbar$ is the resonant Rabi frequency, $\Delta = \omega_0 - \omega$ is the frequency detuning from resonance, and the external field is assumed to vary according to $E = E_0 \cos(\omega t)$. The Eq. (3.10) describes rotation of the Bloch vector around the vector of *constant* angular velocity $\mathbf{W}_0$. We can then achieve a considerable simplification in comparison with the situation described by the original Eq. (3.9). The simplest situation is the case of exact resonance $\Delta = 0$, when rotation of the Bloch vector in the rotating coordinate system and under the action of the external field occurs around axis 1. Then the Bloch vector of a TLS originally in an energy state with lower energy will rotate in the plane 2×3. In this case, after rotation of the Bloch vector through 90°, the system will go from an energy state to a coherent state, and after the next rotation through 90°, it will go to an energy state with a higher energy. Finally, after another rotation through 180°, the system will return to the initial energy state with energy E_1. The described cycle of interaction of the field with the TLS corresponds to self-induced transparency, an effect in which, in the first half of the period, the TLS receives energy from the field, while in the second half of the period, it gives up energy to the field. There are other coherent non-stationary effects in the interaction of radiation with a substance that can be clearly interpreted geometrically with the help of the Bloch vector.

Equations (3.9) and (3.10) do not take into account relaxation processes caused by interaction of the TLS with the environment, so they are true for times shorter than the time of longitudinal and transverse relaxations $t < T_{1,2}$. It is precisely such times that interest us in our considerations. It will be recalled that the longitudinal relaxation time T_1 describes TLS population relaxation, while the transverse relaxation time T_2 characterizes dipole moment relaxation. In terms of the Bloch vector, the first time is connected with relaxation of the third component R_3, and the second is connected with relaxation of the first two components $R_{1,2}$.

3.1.3 System of Equations for Components of the Bloch Vector in Dimensionless Variables

The vector equation for the Bloch vector (3.9) in component form is

$$\frac{dR_1}{dt} = \omega_0 R_2, \tag{3.11}$$

$$\frac{dR_2}{dt} = -\omega_0 R_1 + 2\,\Omega(t)\,R_3, \tag{3.12}$$

$$\frac{dR_3}{dt} = -2\,\Omega(t)\,R_2. \tag{3.13}$$

It will be recalled that here $\Omega(t) = d_0\,E(t)/\hbar$ is the time-dependent "instantaneous" Rabi frequency and $E(t)$ is the electric field strength.

To be specific, we assume hereafter that the optical center in a solid is acted on by a short Gaussian laser pulse, the electric field strength of which is given by

$$E(t) = E_0 \exp\left(-t^2/\Delta t^2\right) \cos(\omega t + \Phi(t)), \tag{3.14}$$

where E_0 is the electric field strength amplitude, Δt is the pulse duration, ω is the carrier frequency, and $\Phi(t)$ is the carrier phase with respect to the pulse envelope, which can be time-dependent (in the case of a chirped pulse).

To render the discussion universal, it is convenient to introduce dimensionless variables: let $\tau = \omega_0\,t$ be the dimensionless time, $\xi = d_0\,E_0/\hbar\,\omega_0$ the dimensionless electric field amplitude, $\eta = \omega_0\,\Delta t$ the dimensionless pulse duration, and $r = \omega/\omega_0$ the dimensionless carrier frequency.

In dimensionless variables the system (3.11)–(3.13) can be represented in the following form, which will be used hereafter for numerical simulation of phototransitions in the TLS:

$$\begin{cases} \dot{R}_1 = R_2, \\ \dot{R}_2 = -\,R_1 + 2\,\xi\,\tilde{E}(\tau)\,R_3, \\ \dot{R}_3 = -\,2\,\xi\,\tilde{E}(\tau)\,R_2, \end{cases} \tag{3.15}$$

where the dot denotes differentiation with respect to the dimensionless time $\tau = \omega_0 t$. The dimensionless electric field strength $\tilde{E}(\tau)$ appearing in (3.15) is determined by

$$\tilde{E}(\tau) = \frac{E(t = \tau/\omega_0)}{E_0}. \tag{3.16}$$

The system (3.15) is true for times shorter than the longitudinal and transverse relaxation times, so it is implied that the pulse duration is short enough, i.e., $\Delta t \ll T_{1,2}$. Generalization to the case of long times presents no problems, but lengthens the formulas. In the general case, the system of differential Eq. (3.15) has no analytical solution, but it is easily solved numerically.

3.2 Photoexcitation in the Perturbation Limit

3.2.1 Harmonic Approximation

If the second component of the Bloch vector is excluded from the system (3.15), we obtain

$$\ddot{R}_1 + R_1 = 2\,\xi\,R_3\,\tilde{E}, \tag{3.17}$$

the left-hand side of which coincides with the equation for a harmonic oscillator. Equation (3.17) describes forced oscillations of a harmonic oscillator corresponding to the TLS if we can assume in the perturbation theory limit that $R_3 \cong 1$. This approximate equality corresponds to *weak excitation* of the TLS when, in the zero approximation, it can be considered that $N_1 \approx 1$ and $N_2 \approx 0$. Then instead of (3.17) we have

$$\ddot{R}_1 + R_1 = \gamma\,\tilde{E}, \tag{3.18}$$

where $\gamma = 2\,\xi = \Omega_0/\omega_0 = const$ is the constant of interaction between the electromagnetic field and the two-level system ($\Omega_0 = d_0\,E_0/2\hbar$ is the resonance Rabi frequency). It will be recalled that, according to (3.6), the first component of the Bloch vector defines the dipole moment of the TLS: $d(t) = R_1\,d_0$. On the other hand, from (3.8), it follows that the amplitude of oscillations of the TLS dipole moment in the absence of external disturbance is related to the third component of the Bloch vector: $d_{amp} = d_0\sqrt{1 - R_3^2}$. In view of the above, we come to the conclusion that the amplitude of *free* oscillations of the first component of the Bloch vector is connected with its third component and defines the TLS population difference (3.4c) in terms of that component.

Thus when the action of the electromagnetic pulse has ended, and when the formula (3.8) is valid, the population of the upper level of the TLS is

$$N_2(t > \Delta t) = \frac{1}{2}\left(1 - \sqrt{1 - R_{1amp}^2}\right). \tag{3.19}$$

This formula was derived using the normalization condition for the TLS, viz., $N_1 + N_2 = 1$, and the determination of the third component of the Bloch vector, viz., $R_3 = N_1 - N_2$. Using (3.19) and the known amplitude of asymptotic oscillations (occurring after termination of the pulse) of the first component, one can find the population of the upper level of the TLS. In the perturbation theory limit, when $R_{1amp} \ll 1$, (3.19) gives

$$N_2(t > \Delta t) \cong \left(R_{1amp}/2\right)^2. \tag{3.20}$$

So when the perturbation theory is valid ($\Omega_0 \ll \omega_0$), solution of (3.18) for times $t > \Delta t$ allows one to find the population of the excited state of the TLS after termination of the electromagnetic pulse.

To obtain the solution of (3.18), we rewrite it in the form

$$\ddot{R}_1 + 2\delta\dot{R}_1 + R_1 = \gamma\tilde{E}(\tau). \tag{3.21}$$

Here we have introduced an infinitesimal damping constant $\delta \to +0$. It is supposed that $\tilde{E}(\tau \to -\infty) = \tilde{E}(\tau \to \infty) = 0$ and $R_1(\tau \to -\infty) = 0$. Introducing the Fourier transforms of the functions involved in (3.21), we obtain

$$R_1(\upsilon) = \gamma\frac{\tilde{E}_\upsilon(\upsilon)}{1 - \upsilon^2 - 2i\upsilon\delta}, \tag{3.22}$$

where $\tilde{E}_\upsilon(\upsilon)$ is the Fourier transform of the field $\tilde{E}(\tau)$. Hence for the required time dependence we find the expression

$$R_1(\tau) = \gamma\int_{-\infty}^{\infty}\frac{\tilde{E}_r(\upsilon)\exp(-i\upsilon\tau)}{1 - \upsilon^2 - 2i\upsilon\delta}\frac{d\upsilon}{2\pi}. \tag{3.23}$$

This can be rewritten in terms of the time integral

$$R_1(\tau) = \gamma\int_{-\infty}^{\infty} G(\tau - \tau')\tilde{E}(\tau')\,d\tau', \tag{3.24}$$

where $G(\tau)$ is the Green function for the harmonic oscillator with damping. Comparing (3.23) and (3.24) and considering the Fourier expansion for the field $\tilde{E}(\tau')$, we obtain the following expression for the Green function of the harmonic oscillator:

$$G(\tau) = \frac{1}{2\pi}\int_{-\infty}^{\infty}\frac{\exp(-i\upsilon\tau)}{1 - \upsilon^2 - 2i\upsilon\delta}\,d\upsilon. \tag{3.25}$$

Thus in the perturbation theory limit, the first component of the Bloch vector represents a harmonic oscillator, so in the case under examination here, the harmonic approximation corresponds to consideration of the TLS within the framework of perturbation theory.

It should be noted that (3.25) coincides with (1.23) if in the latter the replacements $\omega_0 \to 1$ and $\omega' \to \upsilon$ are made. This is connected with the use of dimensionless units in this section. Calculation of the integral on the right-hand side of (3.25) gives

$$G(\tau) = \frac{\Theta(\tau)\,\exp(-\delta\,\tau)}{\sqrt{1-\delta^2}}\,\sin\left[\sqrt{1-\delta^2}\,\tau\right], \tag{3.26}$$

where $\Theta(\tau)$ is the Heaviside step function.

From (3.26) it follows that, in the absence of damping ($\delta = 0$), the Green function of the harmonic oscillator has the very simple form $G(\tau) = \Theta(\tau)\sin(\tau)$. Substituting (3.26) into (3.24), we obtain the expression for the time dependence of forced oscillations of the harmonic oscillator with damping:

$$R_1(\tau) = \frac{\gamma}{\sqrt{1-\delta^2}} \int\limits_0^{\infty} e^{-\delta\tau'}\,\sin\left(\sqrt{1-\delta^2}\,\tau'\right)\tilde{E}(\tau-\tau')\,d\tau'. \tag{3.27}$$

If a new variable $\xi = \tau - \tau'$ is now introduced and it is assumed that $\delta = 0$, (3.27) can be rewritten as

$$R_1(\tau) = \gamma\,[C(\tau)\,\sin\tau \;-\; S(\tau)\,\cos\tau] \tag{3.28}$$

The coefficients $C(\tau)$ and $S(\tau)$ are determined by the integrals

$$C(\tau) = \int\limits_{-\infty}^{\tau} \tilde{E}(\xi)\,\cos\xi\,d\xi, \quad S(\tau) = \int\limits_{-\infty}^{\tau} \tilde{E}(\xi)\,\sin\xi\,d\xi, \tag{3.29}$$

For instants of time when the exciting pulse has already ceased ($\tau \gg \eta$), the upper limit of integration can be replaced by infinity in the formulas (3.29). Then we have

$$C(\tau \gg \eta) \cong C(\infty) = \mathrm{Re}\{\tilde{E}_\upsilon(\upsilon = 1)\},\; S(\tau \gg \eta) \cong S(\infty) = \mathrm{Im}\{\tilde{E}_\upsilon(\upsilon = 1)\}. \tag{3.30}$$

Substituting the expressions (3.30) into (3.28), we arrive at the final formula

$$R_1(\tau \gg \eta) = \gamma\,\left|\tilde{E}_\upsilon(\upsilon = 1)\right|\,\sin\left[\tau - \arg\left(\tilde{E}_\tau(\upsilon = 1)\right)\right], \tag{3.31}$$

where

$$\tilde{E}_{\upsilon}(\upsilon) = \frac{1}{E_0} \int\limits_{-\infty}^{\infty} E\left(t = \frac{\tau}{\omega_0}\right) \exp(i\,\upsilon\tau)\,d\tau \tag{3.32}$$

is the dimensionless Fourier transform of the electric field strength.

Comparing (3.27)–(3.32) and (1.25–1.31), we observe that there is an analogy between the first component of the Bloch vector and the coordinate of the harmonic oscillator. This is not surprising, since the coordinate of the oscillator coincides with its dipole moment up to a constant (the oscillator charge), and the dipole moment of the TLS is equal to the first component of the Bloch vector R_1, up to multiplication by the matrix element d_0 [see (3.6)].

So for the amplitude of oscillations of the first component of the Bloch vector after termination of the exciting pulse, we obtain

$$R_{1amp} = \gamma \left|\tilde{E}_{\upsilon}(\upsilon = 1)\right|. \tag{3.33}$$

Hence, appealing to (3.20) and setting $\gamma = 2\,\xi$, we find

$$N_2(\tau > \eta) = \xi^2 \left|\tilde{E}_{\upsilon}(\upsilon = 1)\right|^2 \tag{3.34}$$

for the population of the upper level of the TLS, where $\xi = d_0\,E_0/\hbar\,\omega_0$, whence the required value is determined by the squared magnitude of the Fourier transform of the field. A condition for validity of (3.34) is fulfilment of the inequation $N_2(\tau > \eta) \ll 1$.

Thus in order to determine the population of the excited state in the perturbation theory limit, one must know the Fourier transform of the electric field strength in the radiation pulse (3.14). Hereafter, as in Chap. 1, we consider two kinds of function $\Phi(t)$: $\Phi(t) = \varphi = const$, of constant CE phase, and $\Phi(t) = \kappa\,t^2$, with chirped pulse (κ is the time chirp).

In the first case, the Fourier transform of the field is

$$E(\omega') = E_0\,\frac{\sqrt{\pi}}{2}\,\Delta t \left\{\exp\left[-i\,\varphi - \frac{(\omega - \omega')^2\,\Delta t^2}{4}\right] + \exp\left[i\,\varphi - \frac{(\omega + \omega')^2\,\Delta t^2}{4}\right]\right\}, \tag{3.35}$$

and in the second case, it is

$$\begin{aligned} E(\omega') = {} & \frac{\sqrt{\pi}\,E_0\,\Delta t}{\sqrt[4]{1+\alpha^2}} \exp\left\{-\frac{\omega^2 + \omega'^2 + 2\,i\,\alpha\,\omega\,\omega'}{\Delta\,\omega^2}\right\} \\ & \times \cos\left\{\frac{1}{2}\,\mathrm{arctg}(\alpha) - \frac{\alpha\,(\omega^2 + \omega'^2) - 2\,i\,\omega\omega'}{\Delta\,\omega^2}\right\}, \end{aligned} \tag{3.36}$$

where $\alpha = \kappa\,\Delta t^2$ is the dimensionless chirp and $\Delta\omega = 2\sqrt{1+\alpha^2}/\Delta t$ is the pulse spectrum width.

Calculating the squared magnitude of the Fourier transform (3.35), changing to dimensionless variables with $\upsilon = \omega'/\omega_0$, and carrying out a number of transformations from the formula (3.34), we obtain

$$N_2(\tau > \eta) \cong \frac{\pi}{4}(\xi\eta)^2 G(r,\eta)\{1 + K(r,\eta)\ \cos(2\varphi)\}, \tag{3.37}$$

for the asymptotic population of the upper level of the TLS excited by a pulse with a variable CE phase. Here we have introduced the function

$$G(r,\eta) = \exp\left[-\frac{\eta^2 (r-1)^2}{2}\right] + \exp\left[-\frac{\eta^2 (r+1)^2}{2}\right], \tag{3.38}$$

which describes the shape of a line of TLS excitation by a short electromagnetic pulse, and the function

$$K(r,\ \eta) = \operatorname{sech}\left(r\eta^2\right), \tag{3.39}$$

which represents the modulation factor for the TLS population with changing CE phase. It should be noted that these functions are equal to the functions (1.36) and (1.37), respectively, if the dimensionless variables r and η are introduced into them. The plots of the functions (3.38) and (3.39) for different values of the parameters are presented in Figs. 1.4 and 1.3 of Chap. 1, respectively. From these figures, we see that, in the limit of long exciting electromagnetic field pulses $\eta \gg 11$, the phase modulation factor of (3.39) is negligible, and in the line shape function, the first summand can be retained on the right-hand side of (3.38).

Thus the dependence of the TLS excitation probability on the CE phase in the perturbation theory limit is true only for subcycle pulses, when $\eta < 6$.

For a chirped pulse (3.36), instead of (3.37)–(3.39), we have

$$N_2(\tau > \eta) = \frac{\pi}{4}\frac{(\xi\eta)^2}{\sqrt{1+\alpha^2}} G'(\alpha,\eta,r)\,[1 + K'(\alpha,\eta,r)\ \cos(f(\alpha,\eta,r))], \tag{3.40}$$

$$G'(\alpha,\eta,r) = \exp\left\{-\frac{\eta^2 (r-1)^2}{2(1+\alpha^2)}\right\} + \exp\left\{-\frac{\eta^2 (r+1)^2}{2(1+\alpha^2)}\right\}, \tag{3.41}$$

$$K'(\alpha,\eta,r) = \operatorname{sech}\left\{\frac{r\eta^2}{1+\alpha^2}\right\}, \tag{3.42}$$

$$f(\alpha,\eta,r) = \frac{\alpha\eta^2 (1+r^2)}{2(1+\alpha^2)} - \operatorname{arctg}(\alpha). \tag{3.43}$$

Note that the formulas (3.37) and (3.40) have a similar structure, and the functions (3.41) and (3.42) coincide with the functions (3.38) and (3.39) for zero chirp $\alpha = 0$.

3.2.2 Rotating Wave Approximation

Within the framework of the rotating wave approximation (3.10) the population of the excited level in resonance ($\omega = \omega_0$) is given by $N_2 = \sin^2(\theta/2)$, where θ is the pulse area defined as $\theta = \int \Omega_0(t)\, dt$, with $\Omega_0(t)$ the resonance Rabi frequency, which takes into account the time dependence of the pulse envelope. For a field with Gaussian envelope (3.14), the pulse area (without considering a chirp) is $\theta = \sqrt{\pi}\, \xi\, \eta$. For a pulse with a frequency chirp, this expression can be changed somewhat to improve conformity with the exact solution and to take into account the dependence of the population of the upper level on the chirp value. As a result, we obtain the following modification of the rotating wave approximation (for $r = 1$):

$$N_2^{RWA} = \sin^2\left(\frac{\sqrt{\pi}\,\cos[0.5\,\mathrm{arctg}(\alpha)]}{2\sqrt[4]{1+\alpha^2}}\,\xi\,\eta\right), \tag{3.44}$$

which takes into account the dependence of the excitation probability on the chirp value α.

In the rotating wave approximation it is also possible to describe excitation of the TLS originally prepared in some superposition state, as specified by the following initial conditions for the expansion coefficients appearing in (3.3):

$$a_1(t=0) = \cos(\psi/2), \quad a_2(t=0) = \sin(\psi/2)\,\exp(-i\,\phi) \tag{3.45}$$

Here the angle ψ (the superposition angle) determines the initial populations and the angle ϕ corresponds to the initial phase of the TLS. The superposition state (3.45) can be prepared from the stationary state under the action of a resonance pulse with area ψ and initial phase ϕ.

The expression for the amplitude of the upper level of the TLS in the rotating wave approximation for a case of exact resonance $\omega = \omega_0$ and an electric field pulse (3.14) with constant phase $\Phi(t) = \varphi$ (the CE phase) has the form

$$\begin{aligned} c_2^{RWA}(t, \omega = \omega_0) &= i\,\cos(\psi/2)\,\sin(\theta(t)/2)\,\exp(-i\,\phi) \\ &\quad + i\sin(\psi/2)\,\cos(\theta(t)/2)\,\exp(-i\,\phi), \end{aligned} \tag{3.46}$$

where

$$\theta(t) = \frac{d_{12}}{\hbar}\int_{-\infty}^{t} E_0(t')\,dt' \tag{3.47}$$

is the "current" value of the pulse area, with $E_0(t')$ the slowly varying amplitude of the electric field strength. From the formula (3.46) for the upper level population, we obtain

$$N_2^{RWA}(t,\,\omega=\omega_0)=0.5\,\{1-\cos\psi\,\cos\theta(t)+\sin\psi\,\sin\theta(t)\,\cos(\varphi-\phi)\}. \tag{3.48}$$

It thus turns out that the upper level population is modulated with a phase equal to the difference between the CE phase φ and the initial phase of the superposition state φ. If the CE phase is equal to the initial phase of the superposition, (3.48) simplifies to $N_2^{RWA}=\sin^2\left(\frac{\psi+\theta}{2}\right)$. Clearly, for $\psi=0$, the last equation reduces to the usual formula, in the rotating wave approximation, for excitation of the TLS from the lower energy state. Using (3.48), the expression for the phase modulation factor in the approximation under consideration may be written in the form

$$K^{RWA}(\omega=\omega_0)=2\frac{N_{2\max}^{RWA}-N_{2\min}^{RWA}}{N_{2\max}^{RWA}+N_{2\min}^{RWA}}=\frac{2\sin\psi\,\sin\theta}{1-\cos\psi\,\cos\theta}. \tag{3.49}$$

Hence it follows in particular that, in the rotating wave approximation, the phase modulation factor of the unexcited TLS, when $\psi=0$, is equal to zero. The calculated curves for the phase modulation factor (3.49) as functions of the angle ψ for different areas of the resonance pulse θ are presented in Fig. 3.4. Using the expression (3.49), it is easy to show that, in the approximation under consideration, a maximum of the phase modulation factor is reached at $\theta=\psi$.

Note that (3.48) implies that, in the limit of small pulse area, the change in the population of the TLS prepared in the superposition state is proportional to the electric field strength $\Delta N_2\propto E_0$, but not to its square, as in the case of the initial energy state, when $\Delta N_2=N_2\propto\theta^2\propto E_0^2$. This is a consequence of interference effects resulting from the fact that the initial TLS state is a superposition.

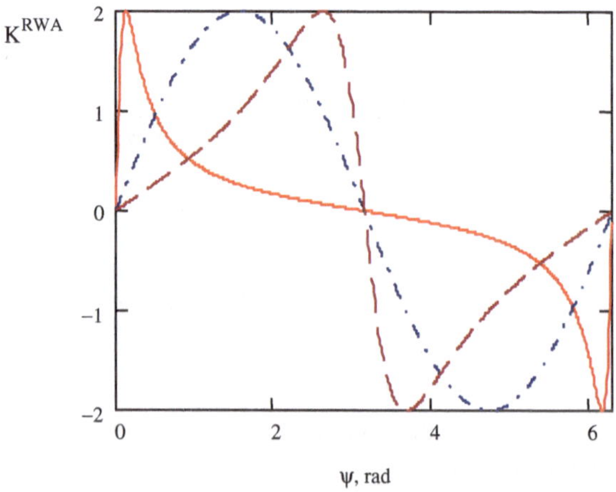

Fig. 3.4 Phase modulation factor as a function of the superposition angle calculated in the rotating wave approximation for excitation of the TLS from the superposition state under the action of a resonance pulse with different areas: *solid curve* $\theta=\pi/24$, *dash-and-dot curve* $\theta=\pi/2$, *dashed curve* $\theta=5\pi/6$

3.3 Phase Effects in Photoexcitation of a Two-Level System in a Strong Field

To describe excitation of the TLS by a high-power electromagnetic pulse, we use the Bloch equations expressed in dimensionless variables as in (3.15). For this purpose, we consider ultrashort Gaussian pulses (3.14) with constant CE phase [2]. Then the system of the Bloch equations can be rewritten as

$$\begin{cases} \dot{R}_1 = R_2, \\ \dot{R}_2 = -R_1 + 2\,\xi \exp\left(-\frac{\tau^2}{\eta^2}\right) \cos(r\tau + \phi_0)\, R_3, \\ \dot{R}_3 = -2\,\xi \exp\left(-\frac{\tau^2}{\eta^2}\right) \cos(r\tau + \phi_0)\, R_2, \end{cases} \tag{3.50}$$

where the dot denotes differentiation with respect to the dimensionless time $\tau = \omega_0\, t$, $\xi = d_0\, E_0/\hbar\, \omega_0$, $\eta = \omega_0\, \Delta t$, $r = \omega/\omega_0$.

3.3.1 *Time Dependence of the Upper Energy Level Population*

We obtain the numerical solution of the system of Eq. (3.50) for the initial condition $\mathbf{R} = (0, 0, 1)$, corresponding to the unexcited state of the TLS before the action of the ultrashort pulse.

To determine the upper level population, we take into account the fact that

$$N_2 = \frac{1 - R_3}{2}. \tag{3.51}$$

The relation (3.51) follows from the determination of (3.4c) and the normalization condition $N_1 + N_2 = 1$.

Figure 3.5 shows the time dependence of the population of the upper level of the two-level system for two different values of the CE phase: $\varphi_0 = \pi/5$, $\pi/2$. The originally unexcited system was excited by a pulse with duration $\eta = 6$ at a high value of the binding force parameter $\xi = 3$. We see that, for the lower value of the CE phase, the asymptotic value of the upper level population is rather high: $N_2(\tau \gg \eta) \approx 0.76$. At the same time, for $\varphi_0 = \pi/2$, the TLS remains practically unexcited after action of the pulse, with $N_2(\tau \gg \eta) \approx 0.05$.

Thus for these values of the parameters, there is efficient phase control of excitation of the TLS due to the change in the CE phase of the Gaussian electric field pulse (3.14).

Strong oscillations of the population during a pulse correspond to the optical nutation of the TLS referred to above in connection with the geometrical interpretation of the evolution of the TLS by means of the Bloch vector.

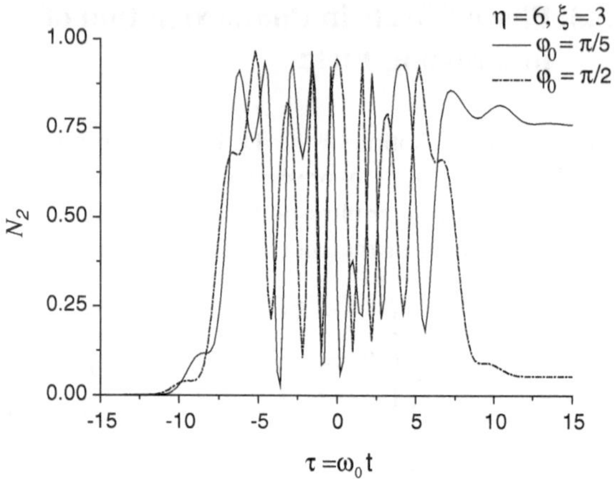

Fig. 3.5 Time dependence of the population of the upper level of the TLS for different values of the CE phase, $\eta = 6, \xi = 3, r = 1$

The calculation shows that, for a lower binding force $\xi = 1$ and the same values of other parameters, the asymptotic populations of the upper level for $\varphi_0 = \pi/5,\ \pi/2$ practically coincide.

In the following, we consider excitation of the TLS by a high-power short pulse with variable phase parameters, using the example of the NV center in diamond (see Fig. 3.1). We carry out the calculation in the "frozen" crystal model, when the phonon subsystem is not explicitly taken into account [3].

3.3.2 *Matrix Element of the Dipole Moment of the Transition*

From the discussion in the previous sections, it follows that, in order to determine the populations of the NV center levels under excitation within the discrete spectrum by short laser pulses, one needs to know the matrix element of the dipole moment of the transition between the levels under consideration $d_0 \equiv d_{21} = \langle 2|\hat{d}|1\rangle$. This matrix element is included in the definition of the dimensionless field amplitude $\xi = d_0 E_0/\hbar\,\omega_0$ which appears in the formulas for the population of the upper energy level of the excited transition.

Thus in order to carry out the calculation, one must use wave functions of the NV center to find the parameter ξ. For this purpose, we use the wave functions in the Slater approximation. The radial Slater functions for the isotropic energy band of a crystal have the form

$$R^{(Sl)}(r, \nu) = \frac{2^{3/2}}{(\nu\, a_n)^{3/2}\sqrt{\Gamma(2\,\nu + 1)}} \left(\frac{2\,r}{\nu\, a_n}\right)^{\nu-1} \exp\left(-\frac{r}{\nu\, a_n}\right), \qquad (3.52)$$

where

$$\nu_n = \frac{1}{K}\sqrt{\frac{m_n/m}{E_{I_n}/Ry}} \tag{3.53}$$

is the effective principal quantum number corresponding to band n, m_n is the effective mass of the corresponding band, K is the dielectric permittivity of the crystal, Ry = 13.6 eV, E_I is the energy of ionization of the impurity center in the specified state, and $a_n = K\hbar^2/m_n e^2$ is the effective Bohr radius.

For the reduced matrix element of the dipole moment, in which there are no angular functions, we have by definition

$$\langle 2\|d\|1\rangle = e\int_0^{\infty} R^{(Sl)}(r,\, \nu_2)\, r\, R^{(Sl)}(r,\, \nu_1)\, r^2\, dr, \tag{3.54}$$

where e is the elementary charge and the radial functions of the initial and final states (3.52) differ by the values of the effective principal quantum number. Substituting the wave functions (3.52) into the (3.54), we find the expression for the reduced matrix element of the dipole moment in the Slater approximation in terms of the effective principal quantum numbers to be

$$\langle 2\|d\|1\rangle = e\, a_n \chi(\nu_1,\, \nu_2), \tag{3.55}$$

where

$$\chi(\nu_1,\, \nu_2) = \frac{2^{\nu_1+\nu_2+1}\, \nu_1^{\nu_2+2}\, \nu_2^{\nu_1+2}\, \Gamma(\nu_1+\nu_2+2)}{(\nu_1+\nu_2)^{\nu_1+\nu_2+2}\sqrt{\nu_1\, \nu_2\, \Gamma(2\,\nu_1+1)\,\Gamma(2\,\nu_2+1)}} \tag{3.56}$$

is a function of the effective principal quantum numbers of the initial and final states and $\Gamma(z)$ is the gamma function.

For transitions $s \to p$, there is a relation between the reduced matrix element of (3.55) and (3.56) and the matrix element $d_0 \equiv d_{21} = \langle 2|\hat{d}|1\rangle$ that takes into account angular integration: $d_0 = \langle 2\|d\|1\rangle/\sqrt{3}$. For the optically allowed transition at the NV center $^3A \leftrightarrow {}^3E$: $d_0 = \langle 2\|d\|1\rangle/\sqrt{2}$, since the statistical weight of the upper energy level (without considering spin) is $g_2 = 2$. In the general case, we have $d_0 = \langle 2\|d\|1\rangle/\sqrt{g_2}$.

The function $\chi(\nu_1,\, \nu_2)$ is plotted in Fig. 3.6 as a function of the parameter ν_2 for different values of the effective principal quantum number of the initial state ν_1.

It is found that the matrix element of the dipole moment increases with the effective principal quantum number ν_1 of the initial state. With increasing effective principal quantum number ν_2 of the excited state, the matrix element $\langle 2\|d\|1\rangle$ decreases monotonically, except for a narrow range of values ν_2 near ν_1.

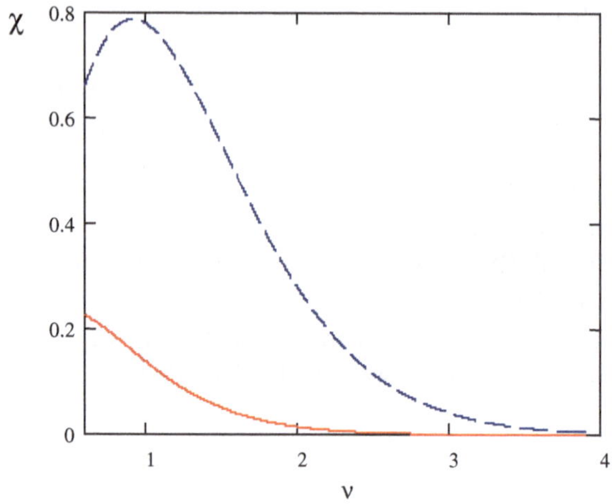

Fig. 3.6 The function $\chi(\nu_1, \nu_2)$ defining the matrix element of the dipole moment in the slater approximation, according to (3.53) and (3.54), constructed in accordance with the effective principal quantum number of the excited state for different values of ν_1: *solid line* $\nu_1 = 0.25$, *dashed line* $\nu_1 = 0.6$

In view of the fact that $\hbar\,\omega_0 = \mathrm{E}_2 - \mathrm{E}_1 = \left(m_n\,e^4/2\,K^2\,\hbar^2\right)\left(\nu_1^{-2} - \nu_2^{-2}\right)$, and given (3.55) and the definition of the parameter $\xi = d_0\,\mathrm{E}_0/\hbar\,\omega_0$, we obtain for this parameter the expression

$$\xi = \frac{(\nu_1\,\nu_2)^2}{\nu_2^2 - \nu_1^2}\,\chi(\nu_1,\,\nu_2)\,\frac{2\,K^3\,\hbar^4}{\sqrt{g_2}\,m_n^2\,e^5}\,\sqrt{8\,\pi\,I/c}, \tag{3.57}$$

in terms of the effective principal quantum numbers of the states and the parameters of the crystal, where $I = c\,E_0^2/8\,\pi$ is the average radiation intensity for the period. It is convenient to rewrite (3.57) in terms of the atomic strength of the electric field $E_a = m^2\,e^5/\hbar^4$:

$$\xi = \frac{2}{\sqrt{g_2}}\,\frac{(\nu_1\,\nu_2)^2}{\nu_2^2 - \nu_1^2}\,\chi(\nu_1,\,\nu_2)\,K^3\left(\frac{m}{m_n}\right)^2\frac{E_0}{E_a}, \tag{3.58}$$

where $\nu_j = \frac{1}{K}\sqrt{\frac{m_n/m}{|\mathrm{E}_j|/Ry}}$ are the effective principal quantum numbers of the initial and final states of the impurity center ($Ry = 13.6$ eV).

To take into account the influence of a medium on the electromagnetic field, one must generally include the multiplier $\zeta = E_{eff}/E$ (the effective field factor) in the formula (3.56). This takes into account the distinction between the effective field E_{eff} acting on the impurity center and the mean field E in the substance. This multiplier is usually treated as an adjustable parameter of the theory. For the NV center, it can be assumed equal to unity.

Thus in view of (3.56), in the Slater approximation, (3.58) gives the expression for the parameter $\xi = d_0\,E_0/\hbar\,\omega_0$ appearing in the formulas for the excitation probability of a bound–bound transition under excitation by a short laser pulse with a modulated phase.

For the NV center, the dependence of the dipole moment of the transition $^3A \to {}^3E$ on the binding energy of the upper level E_2 is presented in Fig. 3.7 in the approximation where the dielectric permittivity is assumed equal to unity, viz., $K = 1$, and the effective mass is assumed equal to the electron mass. Shown as a straight line in Fig. 3.7 is the value of the dipole moment obtained using the experimental value of the Einstein coefficient $A = 7.7 \times 10^7\ \text{s}^{-1}$ for the probability of spontaneous radiation in the transition under consideration. We see that, corresponding to this value of the Einstein coefficient, given the wavelength of radiation in the transition $^3A \to {}^3E$ (637.1 nm), the matrix element of the dipole moment is $d_0 \simeq 3$ a.u. The given values and data of Fig. 3.7 correspond to a binding energy of the upper energy level of 1–2 eV, which agrees with the results of quantum-chemical calculations for the NV center.

So the value of the dimensionless parameter $\xi = d_0 E_0/\hbar\,\omega_0$ required to calculate the population of the upper level of the NV center N_2 for the binding energy $E_2 = 1.2$ eV is $\xi = 43.4(E_0/E_a)$, where $E_a \simeq 5 \times 10^9$ V/cm is the atomic strength of the electric field.

In order to characterize the phase dependence of excitation of the upper level of the NV center under the action of a laser pulse with controlled CE phase, it is convenient to introduce the phase modulation factor by the formula

$$K = \frac{N_2(\varphi = 0) - N_2(\varphi = \pi/2)}{N_2(\varphi = 0) + N_2(\varphi = \pi/2)}, \tag{3.59}$$

where N_2 is the population of the level ^3E after termination of the laser pulse and φ is the CE phase. It is easy to see that, in the harmonic approximation (or the perturbation theory approximation), the phase modulation factor (3.59) coincides with the parameter $K(r, \eta)$ appearing in (3.37).

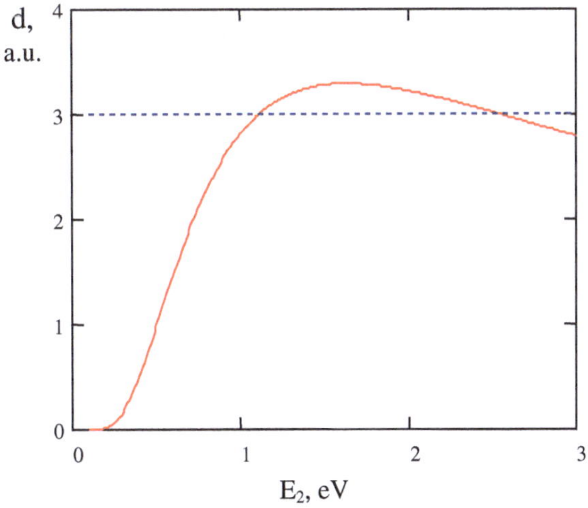

Fig. 3.7 Dependence of the reduced dipole moment of the transition $^3A \to {}^3E$ at the NV center in diamond on the binding energy of the level ^3E

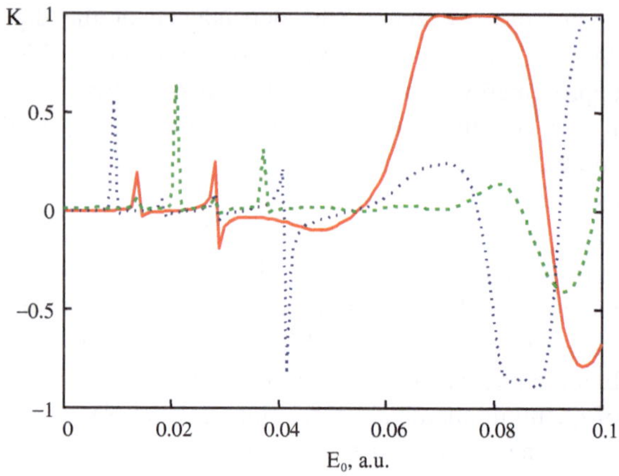

Fig. 3.8 Phase modulation factor for the population of level ^3E of the NV center after the action of the pulse as a function of the electric field amplitude for different pulse durations: *solid curve* single-cycle pulse, *dotted curve* one-and-a-half-cycle pulse, *dashed curve* two-cycle pulse

The results of calculations of the factor (3.59) as a function of the electric field strength are presented in Fig. 3.8 for different pulse durations and $\omega = \omega_0$. These calculations were carried out by numerical solution of the system of equations for the Bloch vector (3.50) and the relation between the population of the upper energy level and the third projection of the Bloch vector (3.51). From Fig. 3.8, we find that, for small values $E_0 < 0.01$ a.u. of the amplitude of the electric field in the pulse, the phase modulation factor is practically equal to zero for all pulse durations shown on the plot.

With growing field amplitude, the phase modulation factor exhibits first narrow dispersion-type extrema, and then wide maxima and minima that are most clearly manifested in the presented range for single-cycle and one-and-a-half-cycle pulses. We see that, for the considered electric field amplitudes, the greatest value of the phase modulation factor occurs for shorter pulses of laser radiation.

Figures 3.9, 3.10, 3.11 show the results from calculations of the population of level ^3E of the NV center as a function of the CE phase for different values of the electric field amplitude and pulse durations.

From Figs. 3.9, 3.10, 3.11, we find that the phase dependence of photoexcitation of the NV center is stronger when the laser pulse is shorter and more intense. In this case the form of the phase dependence differs from the phase dependence in the harmonic limit (3.37). In particular, there is a minimum at $\varphi = \pi$, rather than the maximum indicated by the formula (3.37), and the depth of the minimum depends on the pulse duration.

3.4 Excitation Under the Action of a Chirped Pulse

For excitation of the two-level system by a high-power short pulse, the rotating wave approximation may be found to be inadequate since then the radiation spectrum and the spectral line of the quantum system are considerably broadened,

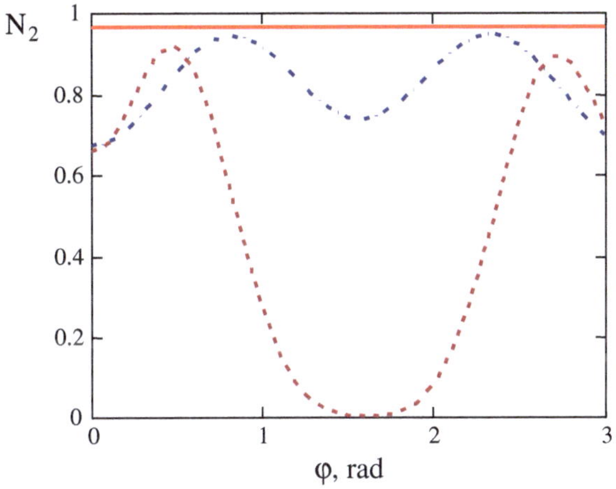

Fig. 3.9 CE phase dependence of the population of level ^3E of the NV center after excitation by a *single-cycle* laser pulse ($\eta = 6$) of different amplitudes: *solid curve* $E_0 = 0.02$ a.u., *dotted curve* $E_0 = 0.04$ a.u., *dashed curve* $E_0 = 0.08$ a.u

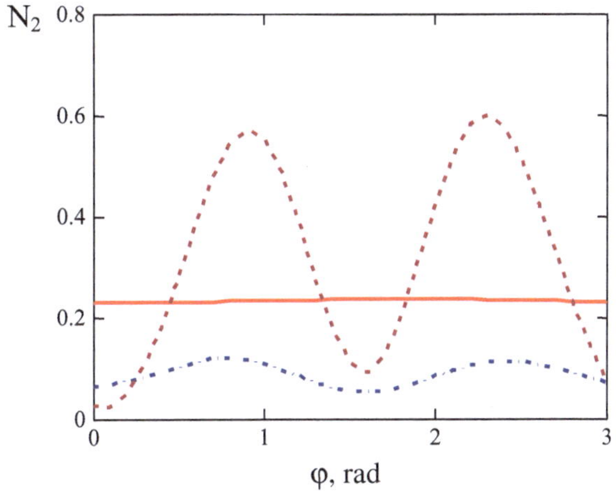

Fig. 3.10 As in Fig. 3.9 but after excitation by a one-and-a-half-cycle pulse ($\eta = 9$)

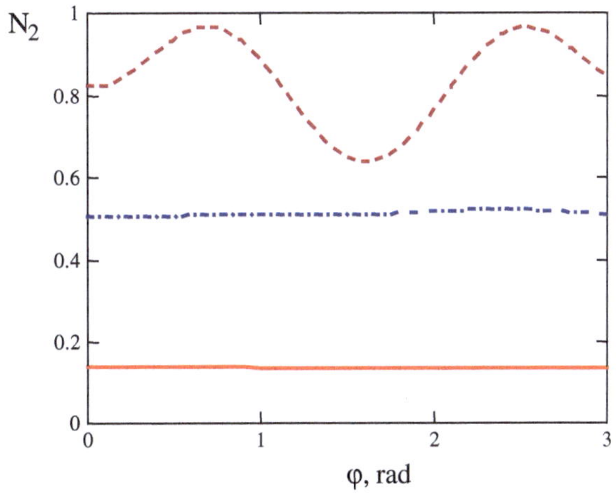

Fig. 3.11 As in Fig. 3.9 but after excitation by a two-cycle pulse ($\eta = 12$)

and the very concept of resonance used to derive the rotating wave approximation formulas becomes inexact.

For arbitrary intensity, duration, and phase dependence of the laser pulse, the population of the excited state can be found by numerical solution of the system of equations for the Bloch vector. For dimensionless variables and action of a chirped pulse on the two-level system, this system has the form [4]

$$\begin{cases} \dot{R}_1 = R_2 \\ \dot{R}_2 = -R_1 + 2\,\xi \exp\left(-\frac{v^2}{\eta^2}\right) \cos\left(rv + \frac{\alpha}{\eta^2} v^2\right) R_3 \\ \dot{R}_3 = -2\,\xi \exp\left(-\frac{v^2}{\eta^2}\right) \cos\left(rv + \frac{\alpha}{\eta^2} v^2\right) R_2 \end{cases} \tag{3.60}$$

where the dot denotes differentiation with respect to the dimensionless time $v = \omega_0 t$. Other parameters are determined by the following equations: $\xi = d_0 E_0/\hbar\,\omega_0$ is the dimensionless field amplitude, $\eta = \omega_0\,\Delta t$ is the dimensionless pulse duration, $r = \omega/\omega_0$ is the dimensionless carrier frequency, and $\alpha = \kappa\,\Delta t^2$ is the dimensionless frequency chirp. It is assumed that the carrier frequency in the chirped pulse varies according to $\omega_c = \omega + \kappa\,t$, where κ is the dimensional frequency chirp.

The system (3.60) is valid for times shorter than the longitudinal and transverse relaxation times considered in this section.

The third projection of the Bloch vector is related to the population of the upper energy level N_2 according to (3.51).

Therefore, solving the system (3.60) numerically with initial condition $\mathbf{R} = (0,\,0,\,1)$, that is, $N_1 = 1, N_2 = 0$, (3.51) can be used to find the population N_2 at times when the action of the laser pulse has ceased and relaxation processes have not yet manifested themselves. We may thus analyze N_2 as a function of the problem parameters, viz., field strength, pulse duration, and frequency chirp.

Figure 3.12 plots the dependence of the population of the excited level of the two-level system on the dimensionless electric field strength parameter $\xi = d_{21}E_0/\hbar\omega_0$ of a resonant single-cycle pulse ($\eta = 6$, $r = 1$) for different values of the dimensionless chirp $\alpha = \kappa\tau^2$, calculated by numerical solution of the system (3.60) and in the modified rotating wave approximation by the formula (3.44). It can be seen that, for zero chirp ($\alpha = 0$), there is a good agreement between the exact solution of the system (3.60) and the result of the rotating wave approximation obtained using (3.44) up to values $\xi \approx 1$ of the strength parameter. And in this case the population as a function of ξ oscillates from zero to one. For the chirp $\alpha = 1$, a characteristic feature of excitation of the two-level system by a chirped pulse is manifested: oscillations of the population as a function of the laser field strength occur from some minimum value $N_{2\,\mathrm{min}} > 0$, increasing to unity with growing chirp.

The calculation shows that variation of the amplitude of oscillations of the population of the upper energy level with changing chirp value does not depend on the pulse duration η. From Fig. 3.12, it can be seen that, for $\alpha = 1$, the generalization (3.44) of the rotating wave approximation taking into account the chirp

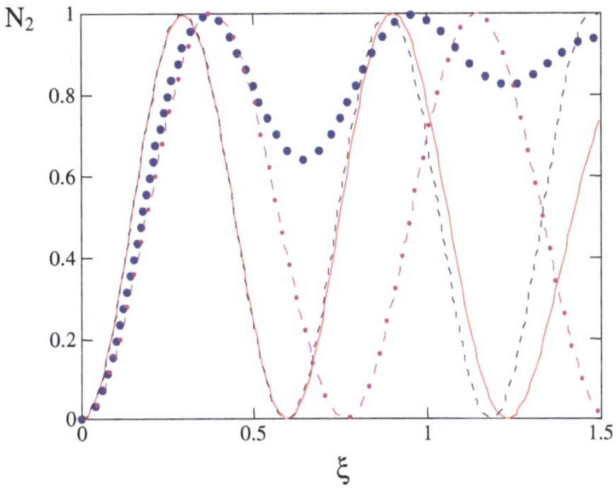

Fig. 3.12 Dependence of the population of the upper level of the two-level system on the electric field strength parameter for a resonant single-cycle pulse ($\eta = 6$, $r = 1$) and two values of the chirp $\alpha = \kappa\tau^2 = 0, 1$: *solid curve* exact solution ($\alpha = 0$), *dashed curve* rotating wave approximation ($\alpha = 0$), *dotted curve* exact solution ($\alpha = 1$), *dash-and-dot curve* rotating wave approximation ($\alpha = 1$)

well describes excitation of the two-level system by a single-cycle pulse up to $\xi_{\max} \approx 0.6$. The value $\xi_{\max}$ decreases with growing pulse duration.

Figure 3.13 shows the dependence of the population of the upper level of the two-level system excited by a two-cycle ($\eta = 12$) resonant ($r = 1$) laser pulse of low intensity $\xi = 0.03$ on the value of the dimensionless chirp $\alpha = \kappa\tau^2$, as calculated by numerical solution of the system of Eq. (3.60) within the framework of the perturbation theory and using (3.40)–(3.43) with the modified rotating wave approximation (3.44).

Fig. 3.13 Population of the upper level of the two-level system excited by a two-cycle resonant laser pulse of low intensity $\xi = 0.03$as a function of the chirp value, calculated by different methods: *solid curve* numerical calculation, *dash-and-dot curve* calculation by the perturbation theory, *dashed curve* calculation in the modified rotating wave approximation

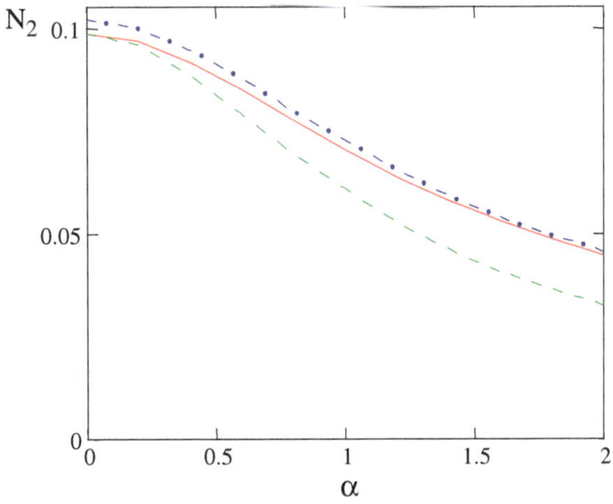

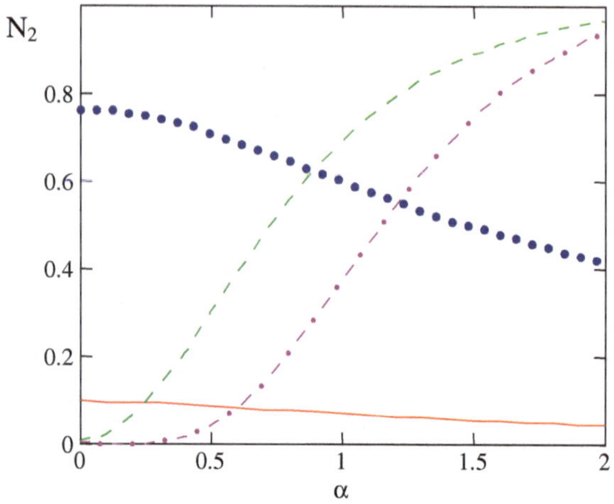

Fig. 3.14 Population of the upper level of the two-level system excited by a two-cycle resonant laser pulse of different intensities as a function of the chirp value: *solid curve* numerical calculation ($\xi = 0.03$), *dotted curve* numerical calculation ($\xi = 0.1$), *dashed curve* numerical calculation ($\xi = 0.3$), *dash-and-dot curve* modified rotating wave approximation ($\xi = 0.3$)

There is good agreement between the results of numerical calculation and the analytical approach within the framework of the perturbation theory. The modified rotating wave approximation, generalized to take into account a chirp, reproduces the exact solution somewhat less well, although the error for these parameter values is insignificant. Note that, in spite of the rather low intensity and the two-cycle nature of the pulse ($\eta = 12$), the dependence of the upper level population on the chirp value is rather appreciable. This contrasts with excitation of the two-level system by a pulse with variable CE phase, when the dependence of the excitation probability on the phase value is vanishingly small under the given conditions.

The calculation shows that the behavior of the probability of excitation of the two-level system as a function of the chirp value varies with changing electric field strength in the laser pulse. This fact is illustrated by the plots of Fig. 3.14, in which the upper level population is shown as a function of the parameter $\alpha = \kappa\tau^2$ for different values of the field strength, calculated by numerical solution of the system of Eq. (3.60). Presented in the same figure is the dependence of $N_2(\alpha, \xi = 0.3)$, calculated using the modified rotating wave approximation and the formula (3.44). From Fig. 3.14, it follows that, with growing electric field strength in the laser pulse, the dependence of the excitation probability on the chirp value changes from decreasing to increasing. For $\xi = 0.3$, there is an especially strong dependence of the upper level population on the chirp parameter α. It should be noted that in this case the modified rotating wave approximation on the whole correctly renders the behavior of the dependence, but numerically understating the value N_2, especially in the region of small values of the parameter α.

Thus there is a range of values of the problem parameters in which efficient control of the probability of excitation of the two-level system is possible by altering the frequency chirp value in the laser pulse. For example, for $\xi = 0.3$, $r = 1$ and a wide interval of values $\eta > 10$ characterizing the pulse duration, the

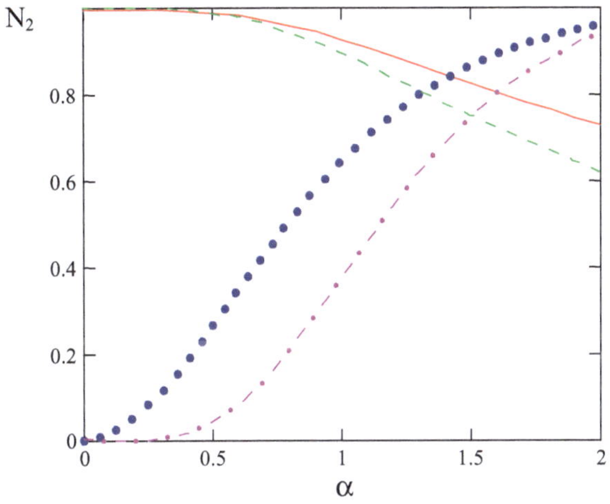

Fig. 3.15 Population of the upper level of the two-level system excited by resonant laser pulses of different durations with the field strength parameter $\xi = 0.3$ as a function of the chirp value: *solid curve* $\eta = 6$, *dotted curve* $\eta = 12$, *dashed curve* $\eta = 6$ (modified rotating wave approximation), *dash-and-dot curve* $\eta = 12$ (modified rotating wave approximation)

probability of system excitation varies from 0.01 to 0.7 when the chirp parameter $\alpha = \kappa\tau^2$ changes from 0 to 1.

The influence of the duration of an exciting pulse on the sensitivity of the upper level population to the chirp value is presented in Fig. 3.15 for the dimensionless strength parameter $\xi = 0.3$ and two values $\eta = 6,\ 12$ of the dimensionless duration. These are the results of numerical calculation based on numerical solution of (3.60) and data obtained using the modified rotating wave approximation with the formula (3.44).

Here we see that, at $\xi = 0.3$ and for a two-cycle laser pulse ($\eta = 12$), the upper level population grows with the chirp value in contrast to the prediction of (3.37)–(3.39) obtained within the framework of the perturbation theory. The analysis shows that, at some values of the electric field strength parameter and pulse duration, for example, for $\xi = 3$ and $\eta = 36$, the population of the upper level of the two-level system is $N_2 \approx 1$ for all chirp values.

So in this section it has been shown that the analytical solution obtained within the framework of the perturbation theory agrees with the numerical result for a low enough value of the dimensionless parameter of the electric field strength $\xi < 0.1$ if the dimensionless pulse duration satisfies the inequation $\eta < 7$. With growing pulse duration η, the maximum strength ξ at which the perturbation theory is applicable decreases.

It has been found that the rotating wave approximation modified to take into account the influence of chirp on the probability of excitation of the two-level system is adequate for moderately low values of the field strength and pulse durations: for a single-cycle pulse the allowed value of the strength parameter is $\xi \leq 0.5$.

We have also shown that there is a wide range of values of the laser radiation electric field strength and pulse durations in which the asymptotic value of the upper energy level population depends essentially on the chirp value. In contrast to the case of laser pulses with variable CE phase, this population is sensitive to chirp in the limit of low intensities for multicycle pulses.

It has been shown that the population of the upper level of the two-level system as a function of the exciting pulse field strength exhibits oscillations with an amplitude that decreases with growing chirp. This contrasts with the case of a non-chirped pulse, where the probability of excitation oscillates as a function of the electric field strength in radiation of unit amplitude.

References

1. Childress, L.I.: Coherent manipulation of single quantum systems in solid state. PhD dissertation, Harvard University Press, Massachusetts (2007)
2. Arustamyan, M.G., Astapenko, V.A.: Phase control of two-level system excitation by short laser pulses. Laser Phys. **18**, 768 (2008)
3. Astapenko, V., Mutafyan, M.: Phase control of excitation of NV centers in diamond by ultra-short laser pulses. Phys. Lett. A **374**, 3701 (2010)
4. Astapenko, V.A., Romadanovskii, M.S.: Excitation of a two-level system by chirped laser pulse. Laser Phys. **19**, 969 (2009)